Dirt

The publisher gratefully acknowledges the generous contribution to this book provided by the General Endowment Fund of the University of California Press Foundation.

Dirt

THE EROSION OF CIVILIZATIONS

David R. Montgomery

UNIVERSITY OF CALIFORNIA PRESS

BERKELEY   LOS ANGELES   LONDON

Also by David R. Montgomery
King of Fish: The Thousand-Year Run of Salmon

University of California Press, one of the most distinguished university presses in the United States, enriches
lives around the world by advancing scholarship in the
humanities, social sciences, and natural sciences. Its activities are supported by the UC Press Foundation and by
philanthropic contributions from individuals and institutions. For more information, visit www.ucpress.edu.

University of California Press
Berkeley and Los Angeles, California

University of California Press, Ltd.
London, England

Library of Congress Cataloging-in-Publication Data

Montgomery, David R., 1961–.
   Dirt : the erosion of civilizations / David R. Mont-
gomery.
       p.   cm.
   Includes bibliographical references and index.
   ISBN-13: 978–0-520–24870–0 (cloth : alk. paper)
   ISBN-10: 0-520–24870–8 (cloth : alk. paper)
     1. Soil science—History.   2. Soils.   3. Soil
erosion.   I. Title.

S590.7.M66   2007
631.4′9—dc22   2006026602

Manufactured in the United States of America

15   14   13   12   11   10   09   08   07
10   9   8   7   6   5   4   3   2

The paper used in this publication meets the minimum
requirements of ANSI/NISO Z39.48–1992 (R 1997) (*Perma-
nence of Paper*).

*For Xena T. Dog, enthusiastic field assistant,
dedicated receptionist, and very best friend—
walk with me forever, sweet girl*

CONTENTS

ACKNOWLEDGMENTS

This book could never have been written without the support of Anne Biklé who, once again, put up with a dining room table covered with endless revisions. Susan Rasmussen chased down obscure historical sources and proved an incredible library sleuth. Polly Freeman, Blake Edgar, and Edith Gladstone provided exceptional editorial input and guidance, and Sam Fleishman was a tremendous help in finding the manuscript a good home. Charles Kiblinger and Harvey Greenberg helped prepare the illustrations. I am also grateful to the Whiteley Center at the University of Washington's Friday Harbor Laboratory for providing the perfect environment to finish the manuscript. I am deeply indebted to the researchers whose work I have relied on in this synthesis and compiled at the end of the book for readers interested in finding the original sources. Naturally, I alone remain responsible for any inadvertent errors and oversights. Finally, in the interest of brevity and narrative I have chosen not to focus on the history and details of the work of the Natural Resources Conservation Service (formerly the Soil Conservation Service), even though its important work remains among the most underappreciated on the planet—and essential to our future.

ONE

Good Old Dirt

What we do to the land, we do to ourselves.

WENDELL BERRY

ON A SUNNY AUGUST DAY IN THE LATE 1990S, I led an expedition up the flank of Mount Pinatubo in the Philippines to survey a river still filled with steaming sand from the massive 1991 eruption. The riverbed jiggled coyly as we trudged upriver under the blazing tropical sun. Suddenly I sank in to my ankles, then my knees, before settling waist deep in hot sand. While my waders began steaming, my graduate students went for their cameras. After properly documenting my predicament, and then negotiating a bit, they pulled me from the mire.

Few things can make you feel as helpless as when the earth gives way beneath your feet. The more you struggle, the deeper you sink. You're going down and there's nothing you can do about it. Even the loose riverbed felt rock solid after that quick dip in boiling quicksand.

Normally we don't think too much about the ground that supports our feet, houses, cities, and farms. Yet even if we usually take it for granted, we know that good soil is not just dirt. When you dig into rich, fresh earth, you can feel the life in it. Fertile soil crumbles and slides right off a shovel. Look closely and you find a whole world of life eating life, a biological orgy recycling the dead back into new life. Healthy soil has an enticing and wholesome aroma—the smell of life itself.

Yet what is dirt? We try to keep it out of sight, out of mind, and outside. We spit on it, denigrate it, and kick it off of our shoes. But in the end, what's more important? Everything comes from it, and everything returns to it. If that doesn't earn dirt a little respect, consider how profoundly soil fertility and soil erosion shaped the course of history.

At the dawn of agricultural civilizations, the 98 percent of people who worked the land supported a small ruling class that oversaw the distribution of food and resources. Today, the less than 1 percent of the U.S. population still working the land feeds the rest of us. Although most people realize how dependent we are on this small cadre of modern farmers, few recognize the fundamental importance of how we treat our dirt for securing the future of our civilization.

Many ancient civilizations indirectly mined soil to fuel their growth as agricultural practices accelerated soil erosion well beyond the pace of soil production. Some figured out how to reinvest in their land and maintain their soil. All depended on an adequate supply of fertile dirt. Despite recognition of the importance of enhancing soil fertility, soil loss contributed to the demise of societies from the first agricultural civilizations to the ancient Greeks and Romans, and later helped spur the rise of European colonialism and the American push westward across North America.

Such problems are not just ancient history. That soil abuse remains a threat to modern society is clear from the plight of environmental refugees driven from the southern plains' Dust Bowl in the 1930s, the African Sahel in the 1970s, and across the Amazon basin today. While the world's population keeps growing, the amount of productive farmland began declining in the 1970s and the supply of cheap fossil fuels used to make synthetic fertilizers will run out later this century. Unless more immediate disasters do us in, how we address the twin problems of soil degradation and accelerated erosion will eventually determine the fate of modern civilization.

In exploring the fundamental role of soil in human history, the key lesson is as simple as it is clear: modern society risks repeating mistakes that hastened the demise of past civilizations. Mortgaging our grandchildren's future by consuming soil faster than it forms, we face the dilemma that sometimes the slowest changes prove most difficult to stop.

For most of recorded history, soil occupied a central place in human cultures. Some of the earliest books were agricultural manuals that passed on knowledge of soils and farming methods. The first of Aristotle's fundamental elements of earth, air, fire, and water, soil is the root of our existence, essential to life on earth. But we treat it as a cheap industrial commodity. Oil is what most of us think of as a strategic material. Yet soil is every bit as important in a longer time frame. Still, who ever thinks about dirt as a strategic resource? In our accelerated modern lives it is easy to forget that fertile soil still provides the foundation for supporting large concentrations of people on our planet.

Geography controls many of the causes of and the problems created by soil erosion. In some regions farming without regard for soil conservation rapidly leads to crippling soil loss. Other regions have quite a supply of fresh dirt to plow through. Few places produce soil fast enough to sustain industrial agriculture over human time scales, let alone over geologic time. Considered globally, we are slowly running out of dirt.

Should we be shocked that we are skinning our planet? Perhaps, but the evidence is everywhere. We see it in brown streams bleeding off construction sites and in sediment-choked rivers downstream from clear-cut forests. We see it where farmers' tractors detour around gullies, where mountain bikes jump deep ruts carved into dirt roads, and where new suburbs and strip malls pave fertile valleys. This problem is no secret. Soil is our most underappreciated, least valued, and yet essential natural resource.

Personally, I'm more interested in asking what it would take to sustain a civilization than in cataloging how various misfortunes can bring down societies. But as a geologist, I know we can read the record previous societies left inscribed in their soils to help determine whether a sustainable society is even possible.

Historians blame many culprits for the demise of once flourishing cultures: disease, deforestation, and climate change to name a few. While each of these factors played varying—and sometimes dominant—roles in different cases, historians and archaeologists rightly tend to dismiss single-bullet theories for the collapse of civilizations. Today's explanations invoke the interplay among economic, environmental, and cultural forces specific to particular regions and points in history. But any society's relationship to its land—how people treat the dirt beneath their feet—is fundamental, literally. Time and again, social and political conflicts undermined societies once there were more people to feed than the land could support. The history of dirt suggests that how people treat their soil can impose a life span on civilizations.

Given that the state of the soil determines what can be grown for how long, preserving the basis for the wealth of future generations requires intergenerational land stewardship. So far, however, few human societies have produced cultures founded on sustaining the soil, even though most discovered ways to enhance soil fertility. Many exhausted their land at a rate commensurate with their level of technological sophistication. We now have the capacity to outpace them. But we also know how not to repeat their example.

Despite substantial progress in soil conservation, the United States Department of Agriculture estimates that millions of tons of topsoil are eroded annually from farmers' fields in the Mississippi River basin. Every second, North America's largest river carries another dump truck's load of topsoil to the Caribbean. Each year, America's farms shed enough soil to fill a pickup truck for every family in the country. This is a phenomenal amount of dirt. But the United States is not the biggest waster of this critical resource. An estimated twenty-four billion tons of soil are lost annually around the world—several tons for each person on the planet. Despite such global losses, soil erodes slowly enough to go largely unnoticed in anyone's lifetime.

Even so, the human cost of soil exhaustion is readily apparent in the history of regions that long ago committed ecological suicide. Legacies of ancient soil degradation continue to consign whole regions to the crushing poverty that comes from wasted land. Consider how the televised images of the sandblasted terrain of modern Iraq just don't square with our notion of the region as the cradle of civilization. Environmental refugees, driven from their homes by the need to find food or productive land on which to grow it, have made headlines for decades. Even when faced with the mute testimony of ruined land, people typically remain unconvinced of the urgent need to conserve dirt. Yet the thin veneer of behavior that defines culture, and even civilization itself, is at risk when people run low on food.

For those of us in developed countries, a quick trip to the grocery store will allay fears of any immediate crisis. Two technological innovations— manipulation of crop genetics and maintenance of soil fertility by chemical fertilizers—made wheat, rice, maize, and barley the dominant plants on earth. These four once-rare plants now grow in giant single-species stands that cover more than half a billion hectares—twice the entire forested area of the United States, including Alaska. But how secure is the foundation of modern industrial agriculture?

Farmers, politicians, and environmental historians have used the term soil exhaustion to describe a wide range of circumstances. Technically, the concept refers to the end state following progressive reduction of crop yields when cultivated land no longer supports an adequate harvest. What defines an adequate harvest could span a wide range of conditions, from the extreme where land can no longer support subsistence farming to where it is simply more profitable to clear new fields instead of working old ones. Consequently, soil exhaustion must be interpreted in the context of social factors, economics, and the availability of new land.

Various social, cultural, and economic forces affect how members of a society treat the land, and how people live on the land, in turn, affects societies. Cultivating a field year after year without effective soil conservation is like running a factory at full tilt without investing in either maintenance or repairs. Good management can improve agricultural soils just as surely as bad management can destroy them. Soil is an intergenerational resource, natural capital that can be used conservatively or squandered. With just a couple feet of soil standing between prosperity and desolation, civilizations that plow through their soil vanish.

As a geomorphologist, I study how topography evolves and how landscapes change through geologic time. My training and experience have taught me to see how the interplay among climate, vegetation, geology, and topography influences soil composition and thickness, thereby establishing the productivity of the land. Understanding how human actions affect the soil is fundamental to sustaining agricultural systems, as well as understanding how we influence our environment and the biological productivity of all terrestrial life. As I've traveled the world studying landscapes and how they evolve, I've come to appreciate the role that a healthy respect for dirt might play in shaping humanity's future.

Viewed broadly, civilizations come and go—they rise, thrive for a while, and fall. Some then eventually rise again. Of course, war, politics, deforestation, and climate change contributed to the societal collapses that punctuate human history. Yet why would so many unrelated civilizations like the Greeks, Romans, and Mayans all last about a thousand years?

Clearly, the reasons behind the development and decline of any particular civilization are complex. While environmental degradation alone did not trigger the outright collapse of these civilizations, the history of their dirt set the stage upon which economics, climate extremes, and war influenced their fate. Rome didn't so much collapse as it crumbled, wearing away as erosion sapped the productivity of its homeland.

In a broad sense, the history of many civilizations follows a common story line. Initially, agriculture in fertile valley bottoms allowed populations to grow to the point where they came to rely on farming sloping land. Geologically rapid erosion of hillslope soils followed when vegetation clearing and sustained tilling exposed bare soil to rainfall and runoff. During subsequent centuries, nutrient depletion or soil loss from increasingly intensive farming stressed local populations as crop yields declined and new land was unavailable. Eventually, soil degradation translated into inadequate agricultural capacity to support a burgeoning population, predisposing whole civilizations to failure. That a similar script appears to apply to small, isolated island societies and extensive, transregional empires suggests a phenomenon of fundamental importance. Soil erosion that outpaced soil formation limited the longevity of civilizations that failed to safeguard the foundation of their prosperity—their soil.

Modern society fosters the notion that technology will provide solutions to just about any problem. But no matter how fervently we believe in its power to improve our lives, technology simply cannot solve the problem of consuming a resource faster than we generate it: someday we will run out of it. The increasingly interconnected world economy and growing population make soil stewardship more important now than anytime in history. Whether economic, political, or military in nature, struggles over the most basic of resources will confront our descendants unless we more prudently manage our dirt.

How much soil it takes to support a human society depends on the size of the population, the innate productivity of the soil, and the methods and technology employed to grow food. Despite the capacity of modern farms to feed enormous numbers of people, a certain amount of fertile dirt must still support each person. This blunt fact makes soil conservation central to the longevity of any civilization.

The capacity of a landscape to support people involves both the physical characteristics of the environment—its soils, climate, and vegetation—and farming technology and methods. A society that approaches the limit of its particular coupled human-environmental system becomes vulnerable to perturbations such as invasions or climate change. Unfortunately, societies that approach their ecological limits are also very often under pressure to maximize immediate harvests to feed their populations, and thereby neglect soil conservation.

Soils provide us with a geological rearview mirror that highlights the importance of good old dirt from ancient civilizations right on through to today's digital society. This history makes it clear that sustaining an industrialized civilization will rely as much on soil conservation and stewardship as on technological innovation. Slowly remodeling the planet without a plan, people now move more dirt around Earth's surface than any other biological or geologic process.

Common sense and hindsight can provide useful perspective on past experience. Civilizations don't disappear overnight. They don't choose to fail. More often they falter and then decline as their soil disappears over generations. Although historians are prone to credit the end of civilizations to discrete events like climate changes, wars, or natural disasters, the effects of soil erosion on ancient societies were profound. Go look for yourself; the story is out there in the dirt.

TWO

Skin of the Earth

We know more about the movement of
celestial bodies than about the soil underfoot.

LEONARDO DA VINCI

CHARLES DARWIN'S LAST AND LEAST-KNOWN BOOK was not particularly controversial. Published a year before he died in 1882, it focused on how earthworms transform dirt and rotting leaves into soil. In this final work Darwin documented a lifetime of what might appear to be trivial observations. Or had he discovered something fundamental about our world—something he felt compelled to spend his last days conveying to posterity? Dismissed by some critics as a curious work of a decaying mind, Darwin's worm book explores how the ground beneath our feet cycles through the bodies of worms and how worms shaped the English countryside.

His own fields provided Darwin's first insights into how worms attain geologic significance. Soon after returning home to England from his voyage around the world, the famous gentleman farmer noticed the resemblance between the stuff worms periodically brought up to the surface and the fine earth that buried a layer of cinders strewn about his meadows years before. Yet since then nothing had happened in these fields, for in them Darwin kept no livestock and grew no crops. How were the cinders that once littered the ground sinking right before his eyes?

About the only explanation that seemed plausible was simply preposterous. Year after year, worms brought small piles of castings up to the surface. Could worms really be plowing his fields? Intrigued, he began investigating whether worms could gradually build up a layer of new soil. Some of his contemporaries thought him crazy—a fool obsessed with the idea that the work of worms could ever amount to anything.

Undeterred, Darwin collected and weighed castings to estimate how much dirt worms moved around the English countryside. His sons helped him examine how fast ancient ruins sank into the ground after they were abandoned. And, most curiously to his friends, he observed the habits of worms kept in jars in his living room, experimenting with their diet and measuring how rapidly they turned leaves and dirt into soil. Darwin eventually concluded that "all the vegetable mould over the whole country has passed many times through, and will again pass many times through, the intestinal canal of worms."[1] It's a pretty big leap to spring from a suspicion about how worms tilled his fields to thinking that they regularly ingested all of England's soil. What led him down this path of unconventional reasoning?

One example in particular stands out among Darwin's observations. When one of his fields was plowed for the last time in 1841, a layer of stones that covered its surface clattered loudly as Darwin's young sons ran down the slope. Yet in 1871, after the field lay fallow for thirty years, a horse could gallop its length and not strike a single stone. What had happened to all those clattering rocks?

Intrigued, Darwin cut a trench across the field. A layer of stones just like those that had once covered the ground lay buried beneath two and a half inches of fine earth. This was just what had happened to the cinders decades before. Over the years, new topsoil built up—a few inches per century—thanks, Darwin suspected, to the efforts of countless worms.

Curious as to whether his fields were unusual, Darwin enlisted his now grown sons to examine how fast the floors and foundations of buildings abandoned centuries before had been buried beneath new soil. Darwin's scouts reported that workmen in Surrey discovered small red tiles typical of Roman villas two and a half feet beneath the ground surface. Coins dating from the second to fourth centuries confirmed that the villa had been abandoned for more than a thousand years. Soil covering the floor of this ruin was six to eleven inches thick, implying it formed at a rate of half an inch to an inch per century. Darwin's fields were not unique.

Observations from other ancient ruins reinforced Darwin's growing belief that worms plowed the English countryside. In 1872 Darwin's son William found that the pavement in the nave of Beaulieu Abbey, which had been destroyed during Henry VIII's war against the Catholic Church, lay six to twelve inches below ground. The ruins of another large Roman villa in Gloucestershire lay undetected for centuries, buried two to three feet under the forest floor until rediscovered by a gamekeeper digging for rabbits. The concrete pavement of the city of Uriconium also lay under almost two feet of soil. These buried ruins confirmed that it took centuries to form a foot of topsoil. But were worms really up to the task?

As Darwin collected and weighed worm castings in a variety of places, he found that each year they brought up ten to twenty tons of earth per acre. Spread across the land in an even layer, all that dirt would pile up a tenth of an inch to a quarter of an inch each year. This was more than enough to explain the burial of Roman ruins and was close to the soil formation rates he'd deduced in what his kids called the stony field. Based on watching and digging in his own fields, excavating the floors of ancient buildings, and directly weighing worm castings, Darwin found that worms played an instrumental role in forming topsoil.

But how did they do it? In the terrariums packed into his cramped living room, Darwin watched worms introduce organic matter into the soil. He counted the huge number of leaves his new pets drew into their burrows as edible insulation. Tearing leaves into small pieces and partially digesting them, worms mixed organic matter with fine earth they had already ingested.

Darwin noticed that in addition to grinding up leaves, worms break small rocks down into mineral soil. When dissecting worm gizzards, he almost always found small rocks and grains of sand. Darwin discovered that the acids in worm stomachs matched humic acids found in soils, and he compared the digestive ability of worms to the ability of plant roots to dissolve even the hardest rocks over time. Worms, it seemed, helped make soil by slowly plowing, breaking up, reworking, and mixing dirt derived from fresh rocks with recycled organic matter.

Darwin discovered that worms not only helped make soil, they helped move it. Prowling his estate after soaking rains he saw how wet castings spread down even the gentlest slopes. He carefully collected, weighed, and compared the mass of castings ejected from worm burrows and found that twice as much material ended up on the downslope side. Material brought up by worms moved an average of two inches downhill. Simply by digging their burrows worms pushed stuff downhill little by little.

Based on his measurements, Darwin calculated that each year a pound of soil would move downslope through each ten-yard-long stretch of a typical English hillslope. He concluded that all across England, a blanket of dirt slowly crept down turf-covered hillsides as an unseen army of worms reworked the soil. Together, English and Scottish worms moved almost half a billion tons of earth each year. Darwin considered worms a major geologic force capable of reshaping the land over millions of years.

Even though his work with worms was, obviously, groundbreaking, Darwin didn't know everything about erosion. He used measurements of the sediment moved by the Mississippi River to calculate that it would take four and a half million years to reduce the Appalachian Mountains to a gentle plain—as long as no uplift occurred. We now know that the Appalachians have been around for over a hundred million years. Geologically dead and no longer rising, they have been eroding away since the time of the dinosaurs. So Darwin massively underestimated the time required to wear down mountains. How could he have been off by so much?

Darwin and his contemporaries didn't know about isostasy—the process through which erosion triggers the uplift of rocks from deep within the earth. The idea didn't enter mainstream geologic thought until decades after his death. Now well accepted, isostasy means that erosion not only removes material, it also draws rock up toward the ground surface to replace most of the lost elevation.

Though at odds with a commonsense understanding of erosion as wearing the world down, isostasy makes sense on a deeper level. Continents are made of relatively light rock that "floats" on Earth's denser mantle. Just like an iceberg at sea, or an ice cube in a glass of water, most of a continent rides down below sea level. Melt off the top of floating ice and what's left rises up and keeps floating. Similarly, the roots of continents can extend down more than fifty miles into the earth before reaching the denser rocks of the mantle. As soil erodes off a landscape, fresh rock rises up to compensate for the mass lost to erosion. The land surface actually drops by only two inches for each foot of rock removed because ten inches of new rock rise to replace every foot of rock stripped off the land. Isostasy provides fresh rock from which to make more soil.

Darwin considered topsoil to be a persistent feature maintained by a balance between soil erosion and disintegration of the underlying rock. He saw topsoil as continuously changing, yet always the same. From watching worms, he learned to see the dynamic nature of Earth's thin blanket of dirt. In this final chapter of his life, Darwin helped open the door for the modern view of soil as the skin of the Earth.

Recognizing their role in making soil, Darwin considered worms to be nature's gardeners.

When we behold a wide, turf-covered expanse, we should remember that its smoothness, on which so much of its beauty depends, is mainly due to all the inequalities having been slowly leveled by worms. It is a marvelous reflection that the whole of the superficial mould over any such expanse has passed, and will again pass, every few years through the bodies of worms. The plough is one of the most ancient and valuable of man's inventions; but long before he existed the land was in fact regularly ploughed, and still continues to be thus ploughed by earthworms. It may be doubted whether there are many other animals which have played so important a part in the history of the world, as have these lowly organised creatures.[2]

Recent studies of the microscopic texture of soils in southeastern Scotland and the Shetland Islands confirm Darwin's suspicions. The topsoil in fields abandoned for several centuries consists almost entirely of worm excrement mixed with rock fragments. As Darwin suspected, it takes worms just a few centuries to thoroughly plow the soil.

Darwin's conception of soil as a dynamic interface between rock and life extended to thinking about how soil thickness reflects local environmental conditions. He described how a thicker soil protects the underlying rocks from worms that penetrate only a few feet deep. Similarly, Darwin noted that the humic acids worms inject into the soil decay before they percolate very far down into the ground. He reasoned that a thick soil would insulate rocks from extreme variations in temperature and the shattering effects of frost and freezing. Soil thickens until it reaches a balance between soil erosion and the rate at which soil-forming processes transform fresh rock into new dirt.

This time Darwin got it right. Soil is a dynamic system that responds to changes in the environment. If more soil is produced than erodes, the soil thickens. As Darwin envisioned, accumulating soil eventually reduces the rate at which new soil forms by burying fresh rock beyond the reach of soil-forming processes. Conversely, stripping the soil off a landscape allows weathering to act directly on bare rock, either leading to faster soil formation or virtually shutting it off, depending on how well plants can colonize the local rock.

Given enough time, soil evolves toward a balance between erosion and the rate at which weathering forms new soil. This promotes development of a characteristic soil thickness for the particular environmental circumstances of a given landscape. Even though a lot of soil may be eroded and

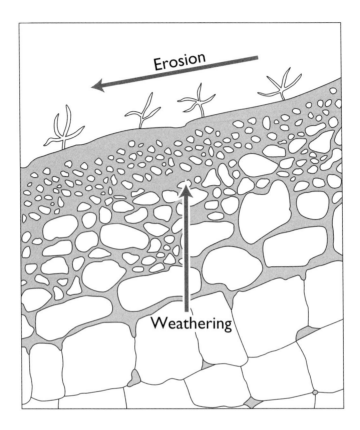

Figure 1. The thickness of hillslope soils represents the balance between their erosion and the weathering of rocks that produces soil.

replaced through weathering of fresh rock, the soil, the landscape, and whole plant communities evolve together because of their mutual interdependence on the balance between soil erosion and soil production.

Such interactions are apparent even in the form of the land itself. Bare angular hillslopes characterize arid regions where the ability of summer thunderstorms to remove soil chronically exceeds soil production. In wetter regions where rates of soil production can keep up with soil erosion, the form of rounded hills reflects soil properties instead of the character of underlying rocks. So arid landscapes where soil forms slowly tend to have angular hillslopes, whereas humid and tropical lands typically have gentle, rolling hills.

Soil not only helps shape the land, it provides a source of essential nutrients in which plants grow and through which oxygen and water are supplied and retained. Acting like a catalyst, good dirt allows plants to capture sunlight and convert solar energy and carbon dioxide into the carbohydrates that power terrestrial life right on up the food chain.

Plants need nitrogen, potassium, phosphorus, and a host of other elements. Some, like calcium or sodium, are common enough that their scarcity does not limit plant growth. Others, like cobalt, are quite rare and yet essential. The processes that create soil also cycle nutrients through ecosystems, and thereby indirectly make the land hospitable to animals as well as plants. Ultimately, the availability of soil nutrients constrains the productivity of terrestrial ecosystems. The whole biological enterprise of life outside the oceans depends on the nutrients soil produces and retains. These circulate through the ecosystem, moving from soil to plants and animals, and then back again into the soil.

The history of life is inextricably related to the history of soil. Early in Earth's history bare rock covered the land. Rainwater infiltrating down into barren ground slowly leached elements out of near-surface materials, transforming rock-forming minerals into clays. Water slowly percolating down through soils redistributed the new clays, forming primitive mineral soils. The world's oldest fossil soil is more than three billion years old, almost as old as the most ancient sedimentary rock and probably land itself. Clay formation appears to have dominated early soil formation; the earliest fossil soils are unusually rich in potassium because there were no plants to remove nutrients from the clays.

Some scientists have proposed that clay minerals even played a key role in the evolution of life by providing highly reactive surfaces that acted as a substrate upon which organic molecules assembled into living organisms. The fossil record of life in marine sediments extends back to about the same time as the oldest soil. Perhaps it is no coincidence that guanine and cytosine (two of the four key bases in DNA) form in clay-rich solutions. Whether or not the breakdown of rocks into clays helped kick-start life, evolution of the earliest soils played a key role in making Earth inhabitable for more complex life.

Four billion years ago Earth's surface temperature was close to boiling. The earliest bacteria were close relatives of those that still carpet Yellowstone's spectacular thermal pools. Fortunately, the growth and development of these heat-loving bacteria increased weathering rates enough to form primitive soils on rocks protected beneath bacterial mats. Their consumption of atmospheric carbon dioxide cooled the planet by 30°C to 45°C—an inverse greenhouse effect. Earth would be virtually uninhabitable were it not for these soil-making bacteria.

The evolution of soils allowed plants to colonize the land. Some 350 million years ago, primitive plants spread up deltas and into coastal valleys where rivers deposited fresh silt eroded off bare highlands. Once plants reached hillsides and roots bound rock fragments and dirt together, primitive soils promoted the breakdown of rocks to form more soil. Respiration by plant roots and soil biota raised carbon dioxide levels ten to a hundred times above atmospheric levels, turning soil water into weak carbonic acid. Consequently, rocks buried beneath vegetation-covered soils decayed much faster than bare rock exposed at the surface. The evolution of plants increased rates of soil formation, which helped create soils better suited to support more plants.

Once organic matter began to enrich soils and support the growth of more plants, a self-reinforcing process resulted in richer soil better suited to grow even more plants. Ever since, organic-rich topsoil has sustained itself by supporting plant communities that supply organic matter back to the soil. Larger and more abundant plants enriched soils with decaying organic matter and supported more animals that also returned nutrients to the soil when they too died. Despite the occasional mass extinction, life and soils symbiotically grew and diversified through climate changes and shifting arrangements of continents.

As soil completes the cycle of life by decomposing and recycling organic matter and regenerating the capacity to support plants, it serves as a filter that cleanses and converts dead stuff into nutrients that feed new life. Soil is the interface between the rock that makes up our planet and the plants and animals that live off sunlight and nutrients leached out of rocks. Plants take carbon directly from the air and water from the soil, but just as in a factory, shortages of essential components limit soil productivity. Three elements—nitrogen, potassium, and phosphorus—usually limit plant growth and control the productivity of whole ecosystems. But in the big picture, soil regulates the transfer of elements from inside the earth to the surrounding atmosphere. Life needs erosion to keep refreshing the soil—just not so fast as to sweep it away altogether.

At the most fundamental level, terrestrial life needs soil—and life plus dirt, in turn, make soil. Darwin estimated that almost four hundred pounds of worms lived in an acre of good English soil. Rich topsoil also harbors microorganisms that help plants get nutrients from organic matter and mineral soil. Billions of microscopic bugs can live in a handful of topsoil; those in a pound of fertile dirt outnumber Earth's human population. That's hard to imagine when you're packed into the Tokyo subway or trying to make your way down the streets of Calcutta or New York City. Yet our reality is built on, and in many ways depends upon, the invisible world of microbes that accelerate the release of nutrients and decay of organic matter, making the land hospitable for plants and therefore people.

Tucked away out of sight, soil-dwelling organisms account for much of the biodiversity of terrestrial ecosystems. Plants supply underground biota with energy by providing organic matter through leaf litter and the decay of dead plants and animals. Soil organisms, in turn, supply plants with nutrients by accelerating rock weathering and the decomposition of organic matter. Unique symbiotic communities of soil-dwelling organisms form under certain plant communities. This means that changes in plant communities lead to changes in the soil biota that can affect soil fertility and, in turn, plant growth.

Along with Darwin's worms, an impressive array of physical and chemical processes help build soil. Burrowing animals—like gophers, termites, and ants—mix broken rock into the soil. Roots pry rocks apart. Falling trees churn up rock fragments and mix them into the soil. Formed under great pressure deep within the earth, rocks expand and crack apart as they near the ground. Big rocks break down into little rocks and eventually into their constituent mineral grains owing to stresses from wetting and drying, freezing and thawing, or heating by wildfires. Some rock-forming minerals, like quartz, are quite resistant to chemical attack. They just break down into smaller and smaller pieces of the same stuff. Other minerals, particularly feldspars and micas, readily weather into clays.

Too small to see individually, clay particles are small enough for dozens to fit on the period at the end of this sentence. All those microscopic clays fit together tightly enough to seal the ground surface and promote runoff of rainwater. Although fresh clay minerals are rich in plant nutrients, once clay absorbs water it holds onto it tenaciously. Clay-rich soils drain slowly and form a thick crust when dry. Far larger, even the smallest sand grains are visible to the naked eye. Sandy soil drains rapidly, making it difficult for plants to grow. Intermediate in size between sand and clay, silt is ideal for growing crops because it retains enough water to nourish plants, yet drains quickly enough to prevent waterlogging. In particular, the mix of clay, silt, and sand referred to as loam makes the ideal agricultural soil because it allows for free air circulation, good drainage, and easy access to plant nutrients.

Clay minerals are peculiar in that they have a phenomenal amount of surface area. There can be as much as two hundred acres of mineral surfaces in half a pound of clay. Like the thin pieces of paper that compose a deck of cards, clay is made up of layered minerals with cations—like potassium, calcium, and magnesium—sandwiched in between silicate sheets. Water that works its way into the clay structure can dissolve cations, contributing to a soil solution rich in plant-essential nutrients.

Fresh clays therefore make for fertile soil, with lots of cations loosely held on mineral surfaces. But as weathering continues, more of the nutrients get leached from a soil as fewer elements remain sandwiched between the silicates. Eventually, few nutrients are left for plants to use. Although clays can also bind soil organic matter, replenishing the stock of essential nutrients like phosphorus and sulfur depends on weathering to liberate new nutrients from fresh rock.

In contrast, most nitrogen enters soils from biological fixation of atmospheric nitrogen. While there is no such thing as a nitrogen-fixing plant, bacteria symbiotic with plant hosts, like clover (to name but one), reduce inert atmospheric nitrogen to biologically active ammonia in root nodules 2–3 mm long. Once incorporated into soil organic matter, nitrogen can circulate from decaying things back into plants as soil microflora secrete enzymes that break down large organic polymers into soluble forms, such as amino acids, that plants can take up and reuse.

How fast soil is produced depends on environmental conditions. In 1941 UC Berkeley professor Hans Jenny proposed that the character of a soil reflected topography, climate, and biology superimposed on the local geology that provides raw materials from which soil comes. Jenny identified five key factors governing soil formation: parent material (rocks), climate, organisms, topography, and time.

The geology of a region controls the kind of soil produced when rocks break down, as they eventually must when exposed at the earth's surface. Granite decomposes into sandy soils. Basalt makes clay-rich soils. Limestone just dissolves away, leaving behind rocky landscapes with thin soils and lots of caves. Some rocks weather rapidly to form thick soils; others resist erosion and only slowly build up thin soils. Because the nutrients available to plants depend on the chemical composition of the soil's parent material, understanding soil formation begins with the rocks from which the soil originates.

Topography also affects the soil. Thin soils with fresh minerals blanket steep slopes in areas where geologic activity raised mountains and continues to refresh slopes. The gentle slopes of geologically quieter landscapes tend to have thicker, more deeply weathered soils.

Climate strongly influences soil formation. High rainfall rates and hot temperatures favor chemical weathering and the conversion of rock-forming minerals into clays. Cold climates accelerate the mechanical breakdown of rocks into small pieces through expansion and contraction during freeze-thaw cycles. At the same time, cold temperatures retard chemical weathering. So alpine and polar soils tend to have lots of fresh mineral surfaces that can yield new nutrients, whereas tropical soils tend to make poor agricultural soils because they consist of highly weathered clays leached of nutrients.

Temperature and rainfall primarily control the plant communities that characterize different ecosystems. At high latitudes, perpetually frozen ground can support only the low scrub of arctic tundra. Moderate temperatures and rainfall in temperate latitudes support forests that produce organic-rich soils by dropping their leaves to rot on the ground. Drier grassland soils that support a lot of microbial activity receive organic matter both from the recycling of dead roots and leaves and from the manure of grazing animals. Arid environments typically have thin rocky soils with sparse vegetation. Hot temperatures and high rainfall near the equator produce lush rainforests growing on leached-out soils by recycling nutrients inherited from weathering and recycled from decaying vegetation. In this way, global climate zones set the template upon which soils and vegetation communities evolved.

Differences in geology and climate make soils in different regions more or less capable of sustained agriculture. In particular, the abundant rainfall and high weathering rates on the gentle slopes of many tropical landscapes mean that after enough time, rainfall seeping into the ground leaches out almost all of the nutrients from both the soil and the weathered rocks beneath the soil. Once this happens, the lush vegetation essentially feeds on itself, retaining and recycling nutrients inherited from rocks weathered long ago. As most of the nutrients in these areas reside not in the soil but in the plants themselves, once the native vegetation disappears, so does the productive capacity of the soil. Often too few nutrients remain to support either crops or livestock within decades of deforestation. Nutrient-poor tropical soils illustrate the general rule that life depends on recycling past life.

Humans have not yet described all the species present in any natural soil. Yet soils and the biota that inhabit them provide clean drinking water, recycle dead materials into new life, facilitate the delivery of nutrients to plants, store carbon, and even remediate wastes and pollutants—as well as produce almost all of our food.

Out of sight and out of mind, soil-dwelling organisms can be greatly influenced by agricultural practices. Tilling the soil can kill large soil-dwelling organisms, and reduce the number of earthworms. Pesticides can exterminate microbes and microfauna. Conventional short-rotation, single-crop farming can reduce the diversity, abundance, and activity of beneficial soil fauna, and indirectly encourage proliferation of soilborne viruses, pathogens, and crop-eating insects. Generally, so-called alternative agricultural systems tend to better retain soil-dwelling organisms that enhance soil fertility.

Like soil formation, soil erosion rates depend on soil properties inherited from the parent material (rocks), and the local climate, organisms, and topography. A combination of textural properties determines a soil's ability to resist erosion: its particular mix of silt, sand, or clay, and binding properties from aggregation with soil organic matter. Higher organic matter content inhibits erosion because soil organic matter binds soil particles together, generating aggregates that resist erosion. A region's climate influences erosion rates through how much precipitation falls and whether it flows off the land as rivers or glaciers. Topography matters as well; all other things being equal, steeper slopes erode faster than gentle slopes. However, greater rainfall not only generates more runoff, and therefore more erosion, it also promotes plant cover that protects the soil from erosion. This basic trade-off means that the amount of rainfall does not simply dictate the pace of soil erosion. Wind can be a dominant erosion process in arid environments or on bare disturbed soil, like agricultural fields. Biological processes, whether Darwin's worms or human activities such as plowing, also gradually move soil downslope.

Although different types of erosional processes are more or less important in different places, a few tend to dominate. When rain falls onto the ground it either sinks into the soil or runs off over it; greater runoff leads to more erosion. Where enough runoff accumulates, flowing sheets of water can pick up and transport soil, carving small channels, called rills, which collect into larger, more erosive gullies—the name for incised channels large enough that they cannot be plowed over. On steep slopes, intense or sustained rainfall can saturate soil enough to trigger landsliding.

Wind can pick up and erode dry soil with sparse vegetation cover. While many of these processes operate in a landscape, the dominant process varies with the topography and climate.

In the 1950s soil erosion researchers began seeking a general equation to explain soil loss. Combining data from erosion research stations they showed that soil erosion, like soil production, is controlled by the nature of the soil, the local climate, the topography, and the nature and condition of the vegetation. In particular, rates of soil erosion are also strongly influenced by the slope of the land and by agricultural practices. Generally, steeper slopes, greater rainfall, and sparser vegetation lead to more erosion.

Plants and the litter they produce protect the ground from the direct impact of raindrops as well as the erosive action of flowing water. When bare soil is exposed to rain, the blast from each incoming raindrop sends dirt downslope. Intense rainfall that triggers rapid topsoil erosion exposes deeper, denser soil that absorbs water less quickly and therefore produces more runoff. This, in turn, increases the erosive power of the water flowing over the ground surface. Some soils are incredibly sensitive to this positive feedback that can rapidly strip topsoil from bare exposed ground.

Below the surface, extensive networks of roots link plants and stabilize the topography. In a closed canopy forest, roots from individual trees intertwine in a living fabric that helps bind soil onto slopes. Conversely, steep slopes tend to erode rapidly when stripped of forest cover.

Soil scientists use a simple system to describe different soil layers—literally an ABC of dirt. The partially decomposed organic matter found at the ground surface is called the O horizon. This organic layer, whose thickness varies with vegetation and climate, typically consists of leaves, twigs, and other plant material on top of the mineral soil. The organic horizon may be missing altogether in arid regions with sparse vegetation, whereas in thick tropical jungles the O horizon holds most soil nutrients.

Below the organic horizon lies the A horizon, the nutrient-rich zone of decomposed organic matter mixed with mineral soil. Dark, organic-rich A horizons at or near the ground surface are what we normally think of as dirt. Topsoil formed by the loose O and A horizons erodes easily if exposed to rainfall, runoff, or high winds.

The next horizon down, the B horizon, is generally thicker than the topsoil, but less fertile due to lower organic content. Often referred to as subsoil, the B horizon gradually accumulates clays and cations carried down into the soil. The weathered rock below the B horizon is called the C horizon.

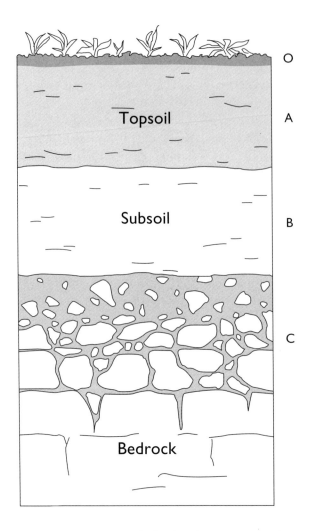

Figure 2. Over time, soils develop distinctive topsoil and subsoil horizons above weathered rock.

Concentrated organic matter and nutrients make soils with well-developed A horizons the most fertile. In topsoil, a favorable balance of water, heat, and soil gases fosters rapid plant growth. Conversely, typical subsoils have excessive accumulations of clay that are hard for plant roots to penetrate, low pH that inhibits crop growth, or cementlike hardpan layers enriched in iron, aluminum, or calcium. Soils that lose their topsoil generally are less productive, as most B horizons are far less fertile than the topsoil.

Combinations of soil horizons, their thickness, and composition vary widely for soils developed under different conditions and over different lengths of time. There are some twenty thousand specific soil types recognized in the United States. Despite such variety, most soil profiles are about one to three feet thick.

Soil truly is the skin of the earth—the frontier between geology and biology. Within its few feet, soil accounts for a bit more than a ten millionth of our planet's 6,380 km radius. By contrast, human skin is less than a tenth of an inch thick, a little less than a thousandth of the height of the average person. Proportionally, Earth's skin is a much thinner and more fragile layer than human skin. Unlike our protective skin, soil acts as a destructive blanket that breaks down rocks. Over geologic time, the balance between soil production and erosion allows life to live off a thin crust of weathered rock.

The global geography of soil makes a few key regions particularly well suited to sustaining intensive agriculture. Most of the planet has poor soils that are difficult to farm, or are vulnerable to rapid erosion if cleared and tilled. Globally, temperate grassland soils are the most important to agriculture because they are incredibly fertile, with thick, organic-rich A horizons. Deep and readily tilled, these soils underlie the great grain-producing regions of the world.

A civilization can persist only as long as it retains enough productive soil to feed its people. A landscape's soil budget is just like a family budget, with income, expenses, and savings. You can live off your savings for only so long before you run out of money. A society can remain solvent by drawing off just the interest from nature's savings account—losing soil only as fast as it forms. But if erosion exceeds soil production, then soil loss will eventually consume the principal. Depending on the erosion rate, thick soil can be mined for centuries before running out; thin soils can disappear far more rapidly.

Instead of the year-round plant cover typical of most native vegetation communities, crops shield agricultural fields for just part of the year, exposing bare soil to wind and rain and resulting in more erosion than would occur under native vegetation. Bare slopes also produce more runoff, and can erode as much as a hundred to a thousand times faster than comparable vegetation-covered soil. Different types of conventional cropping systems result in soil erosion many times faster than under grass or forest.

In addition, soil organic matter declines under continuous cultivation as it oxidizes when exposed to air. Thus, because high organic matter content can as much as double erosion resistance, soils generally become more erodible the longer they are plowed.

Conventional agriculture typically increases soil erosion to well above natural rates, resulting in a fundamental problem. The United States Department of Agriculture estimates that it takes five hundred years to produce an inch of topsoil. Darwin thought English worms did a little better, making an inch of topsoil in a century or two. While soil formation rates vary in different regions, accelerated soil erosion can remove many centuries of accumulated soil in less than a decade. Earth's thin soil mantle is essential to the health of life on this planet, yet we are gradually stripping it off—literally skinning our planet.

But agricultural practices can also retard erosion. Terracing steep fields can reduce soil erosion by 80 to 90 percent by turning slopes into a series of relatively flat surfaces separated by reinforced steps. No-till methods minimize direct disturbance of the soil. Leaving crop residue at the ground surface instead of plowing it under acts as mulch, helping to retain moisture and retard erosion. Interplanting crops can provide more complete ground cover and retard erosion. None of these alternative practices are new ideas. But the growing adoption of them is.

Over decades of study, agronomists have developed ways to estimate soil loss for different environmental conditions and under different agricultural practices relative to standardized plots. Despite half a century of first-rate research, rates of soil erosion remain difficult to predict; they vary substantially both from year to year and across a landscape. Decades of hard-to-collect measurements are needed to get representative estimates that sample the effect of rare large storms and integrate the effects of common showers. The resulting uncertainty as to the relative magnitude of modern erosion rates has contributed to controversy in the last few decades over whether soil loss is a serious problem. Whether it is depends on the ratio of soil erosion to soil production, and even less is known about rates of soil formation than about rates of soil erosion.

Skeptics discount concern over erosion rates measured from small areas or experimental plots and extrapolated using models to the rest of the landscape. They rightly argue that real data on soil erosion rates are hard to come by, locally variable, and require decades of sustained effort to get. In their view, we might as well be guessing an answer. Moreover, only sparse data on soil production rates have been available until the last few decades. Yet the available data do show that conventional agricultural methods accelerate erosion well beyond soil production—the question is by how much. This leaves the issue in a position not unlike global warming—while academics argue over the details, vested interests stake out positions to defend behind smokescreens of uncertainty.

Still, even with our technological prowess, we need productive soil to grow food and to support plants we depend on—and our descendants will too. On the hillslopes that support much of our modern agriculture, soil conservation is an uphill battle. But there are some places where hydrology and geology favor long-term agriculture—the fertile river valleys along which civilizations first arose.

THREE

Rivers of Life

Egypt is the gift of the Nile.

HERODOTUS

FOUNDATIONAL TEXTS OF WESTERN RELIGIONS acknowledge the fundamental relationship between humanity and the soil. The Hebrew name of the first man, Adam, is derived from the word *adama,* which means earth, or soil. Because the name of Adam's wife, Eve, is a translation of *hava,* Hebrew for "living," the union of the soil and life linguistically frames the biblical story of creation. God created the earth—Adam—and life—Eve—sprang from the soil—Adam's rib. The Koran too alludes to humanity's relation to the soil. "Do they not travel through the earth and see what was the end of those before them? . . . They tilled the soil and populated it in greater numbers . . . to their own destruction" (Sura 30:9). Even the roots of Western language reflect humanity's dependence on soil. The Latin word for human, *homo,* is derived from *humus,* Latin for living soil.

The image of a lush garden of Eden hardly portrays the Middle East today. Yet life for the region's Ice Age inhabitants was less harsh than along the great northern ice sheets. As the ice retreated after the peak of the last glaciation, game was plentiful and wild stands of wheat and barley could be harvested to supplement the hunt. Are vague cultural memories of a prior climate and environment recorded in the story of the garden from which humanity was ejected before the rise of civilization?

Regardless of how we view such things, the changing climate of the last two million years rearranged the world's ecosystems time and again. The

Ice Age was not a single event. More than twenty major glaciations repeatedly buried North America and Europe under ice, defining what geologists call the Quaternary—the fourth era of geologic time.

At the peak of the most recent glaciation, roughly 20,000 years ago, glaciers covered almost a third of Earth's land surface. Outside of the tropics even unglaciated areas experienced extreme environmental changes. Human populations either adapted, died out, or moved on as their hunting and foraging grounds shifted around the world.

Each time Europe froze, North Africa dried, becoming an uninhabitable sand sea. Naturally, people left. Some migrated south back into Africa. Others ventured east to Asia or into southern Europe as periodic climate upheavals launched the great human migrations that eventually circled the world.

Judged by the fossil evidence, *Homo erectus* walked out of Africa and ventured east across Asia, sticking to tropical and temperate latitudes about two million years ago just after the start of the glacial era. Fossil and DNA evidence indicates that the initial separation of Neanderthals from the ancestors of genetically modern humans occurred at least 300,000 years ago—about the time Neanderthals arrived in Europe and western Asia. After successfully adapting to the glacial climate of northwestern Eurasia, Neanderthals disappeared as a new wave of genetically modern humans spread from Africa through the Middle East around 45,000 years ago and across Europe by at least 35,000 years ago. People continued spreading out across the world when the Northern Hemisphere's great ice sheets once again plowed southward, rearranging the environments of Europe, northern Africa, and the Middle East.

During the most recent glaciation, large herds of reindeer, mammoth, wooly rhinoceroses, and giant elk roamed Europe's frozen plains. Ice covered Scandinavia, the Baltic coast, northern Britain, and most of Ireland. Treeless tundra stretched from France through Germany, on to Poland and across Russia. European forests shrank to a narrow fringe around the Mediterranean. Early Europeans lived through this frozen time by following and culling herds of large animals. Some of these species, notably wooly rhinos and giant elk, did not survive the transition to the postglacial climate.

Extreme environmental shifts also isolated human populations and helped differentiate people into the distinct appearances we know today as races. Skin shields our bodies and critical organs from ultraviolet radiation. But skin must also pass enough sunlight to support production of the vitamin D needed to make healthy bones. As our ancestors spread around the globe, these opposing pressures colored the skin of people in different regions. The dominant need for UV protection favored dark skin in the tropics; the need for vitamin D favored lighter skin in the northern latitudes.

Technological innovation played a key role in the spread and adaptation of people to new environments. Roughly 30,000 years ago, immediately before the last glaciation, the development of thin, sharp stone tools ushered in a major technological revolution. Then, about 23,000 years ago, just before the last glacial maximum, the art of hunting changed radically as the bow and arrow began to replace spears. Development of eyed needles allowed the production of hoods, gloves, and mittens from wooly animal hides. Finally equipped to endure the long winter of another glacial era, central Asian hunters began following large game across the grassy steppe west into Europe, or east into Siberia and on to North America.

Unglaciated areas also experienced dramatic shifts in vegetation as the planet cooled and warmed during glacial and interglacial times. Long before the last glacial advance, people around the world burned forest patches to maintain forage for game or to favor edible plants. Shaping their world to suit their needs, our hunting and gathering ancestors were not passive inhabitants of the landscape. Despite their active manipulation, small human populations and mobile lifestyles left little discernable impact on natural ecosystems.

Transitions from a glacial to interglacial world occurred many times during the last two million years. Through all but the most recent glaciation, people moved along with their environment rather than staying put and adapting to a new ecosystem. Then, after living on the move for more than a million years, they started to settle down and become farmers. What was so different when the glaciers melted this last time that caused people to adopt a new lifestyle?

Several explanations have been offered to account for this radical change. Some argue that the shift from a cool, wet glacial climate to less hospitable conditions put an environmental squeeze on early people in the Middle East. In this view, hunters began growing plants in order to survive when the climate warmed and herds of wild game dwindled. Others argue that agriculture evolved in response to an inevitable process of cultural evolution without any specific environmental forcing. Whatever the reasons, agriculture developed independently in Mesopotamia, northern China, and Mesoamerica.

For much of the last century, theories for the origin of agriculture emphasized the competing oasis and cultural evolution hypotheses. The oasis hypothesis held that the postglacial drying of the Middle East restricted edible plants, people, and other animals to well-watered flood-plains. This forced proximity bred familiarity, which eventually led to domestication. In contrast, the cultural evolution hypothesis holds that regional environmental change was unimportant in the gradual adoption of agriculture through an inevitable progression of social development. Unfortunately, neither hypothesis provides satisfying answers for why agriculture arose when and where it did.

A fundamental problem with the oasis theory is that the wild ancestors of our modern grains came to the Middle East from northern Africa at the end of the last glaciation. This means that the variety of food resources available to people in the Middle East was expanding at the time that agriculture arose—the opposite of the oasis theory. So the story cannot be as simple as the idea that people, plants, and animals crowded into shrinking oases as the countryside dried. And because only certain people in the Middle East adopted agriculture, the cultural adaptation hypothesis falls short. Agriculture was not simply an inevitable stage on the road from hunting and gathering to more advanced societies.

The transition to an agricultural society was a remarkable and puzzling behavioral adaptation. After the peak of the last glaciation, people herded gazelles in Syria and Israel. Subsisting on these herds required less effort than planting, weeding, and tending domesticated crops. Similarly, in Central America several hours spent gathering wild corn could provide food for a week. If agriculture was more difficult and time-consuming than hunting and gathering, why did people take it up in the first place?

Increasing population density provides an attractive explanation for the origin and spread of agriculture. When hunting and gathering groups grew beyond the capacity of their territory to support them, part of the group would split off and move to new territory. Once there was no more productive territory to colonize, growing populations developed more intensive (and time-consuming) ways to extract a living from their environment. Such pressures favored groups that could produce food themselves to get more out of the land. In this view, agriculture can be understood as a natural behavioral response to increasing population.

Modern studies have shown that wild strains of wheat and barley can be readily cultivated with simple methods. Although this ease of cultivation suggests that agriculture could have originated many times in many places, genetic analyses show that modern strains of wheat, peas, and lentils all came from a small sample of wild varieties. Domestication of plants fundamental to our modern diet occurred in just a few places and times when people began to more intensively exploit what had until then been secondary resources.

The earliest known semiagricultural people lived on the slopes of the Zagros Mountains between Iraq and Iran about 11,000 to 9000 BC (or thirteen thousand to eleven thousand years ago). Surviving by hunting gazelles, sheep, and goats and gathering wild cereals and legumes, these people occupied small villages but made extensive use of seasonal hunting camps and caves. By 7500 BC herding and cultivation replaced hunting and gathering as the mainstay of their diet and settled villages of up to twenty-five households kept sheep and goats and grew wheat, barley, and peas. By then hunting accounted for only about 5 percent of their food. Why the big change, and why then and there?

The earliest evidence for systematic cultivation of grains comes from Abu Hureyra in the headwaters of the Euphrates River in modern Syria. The archaeological record from this site shows that cultivation began in response to a period when the drier conditions of glacial times abruptly returned after thousands of years of climatic amelioration. Abu Hureyra provides a unique record of the transition from the hunter-gathering lifestyle of the last glacial era to cereal-based agriculture. Moreover, evidence from the site helps explain why people adopted the labor-intensive business of agriculture. They were forced into it.

As glaciation ended, the Levant gradually warmed and received increasing rainfall. From about 13,000 to 11,000 BC open oak forest gradually replaced the grasslands of the glacial steppe. A core drilled from the bed of Lake Huleh in northeastern Israel shows that tree pollen increased from a fifth to three-quarters of all the pollen during this period. Abundant game and wild grains (especially rye and wheat) made for an edenic landscape with few people and lots of resources. Sedentary communities of hunter-gatherers began to take root in locations where resources were particularly abundant.

Then the world's climate reverted to almost full glacial conditions for a thousand years, from about 10,000 to 9000 BC, a period known as the Younger Dryas. Arboreal pollen dropped back to less than a quarter of the total amount of pollen, indicating a sharp decline in precipitation and a return to the steppelike conditions of the glacial climate. The forest retreated northward, away from the world's first settled community.

Abu Hureyra sat on a low promontory overlooking the Euphrates Valley, about 180 miles northeast of Damascus. Plant debris excavated from the site records the transition from foraging for a wide variety of wild plants to cultivation of a few crops by the end of the Younger Dryas. The earliest plant remains associated with settlement of the site include more than one hundred species of seeds and fruits from the marshes and forest of the Euphrates Valley. Abundant animal bones reveal substantial reliance on hunting, especially gazelles. Moreover, the site was occupied year-round. The people of Abu Hureyra were not nomadic hunter-gatherers. They permanently inhabited a defined territory around their village. A couple hundred people occupied Abu Hureyra by the time that the Younger Dryas ushered in a thousand years of cold, dry weather that dramatically altered plant and animal resources. Fruits and seeds of drought-sensitive plants disappeared from the diet. Wild lentils and legumes harvested from nearby woodland also disappeared. As eden dried out, food became scarce.

Why didn't they just move? Probably because Abu Hureyra was already one of the region's best sites. Surrounding areas experienced similar changes and offered even less sustenance. Besides, other people already occupied the next best land. People with rapidly disappearing food supplies usually do not welcome new neighbors. The people of Abu Hureyra had no place to go.

Out of options, they began to cultivate wild varieties of rye and wheat that survived the transition to a colder, more arid climate. Of the plants that survived, only cereals could be cultivated to produce food capable of storage for use throughout the year. Despite the worsening aridity, seeds of drought-intolerant weeds typical of agricultural fields increased dramatically during the Younger Dryas. At first, wild cereals were cultivated on hillsides using rain-fed agriculture. Within a few centuries domesticated varieties of rye appeared in the fields, as did legumes such as lentils.

The switch to cultivation required more time and energy to produce a calorie of food. It is not something that would have been undertaken lightly. The sedentary style of hunting and gathering practiced by the early inhabitants of Abu Hureyra left them susceptible to declining food availability as the climate changed. Once wild food sources were fully exploited the population was vulnerable to seasonal shortages brought on by increasing aridity. Begun out of desperation, agriculture expanded to include other crops such as barley and peas as the climate improved after the Younger Dryas ended. Settlement around Abu Hureyra grew rapidly in the

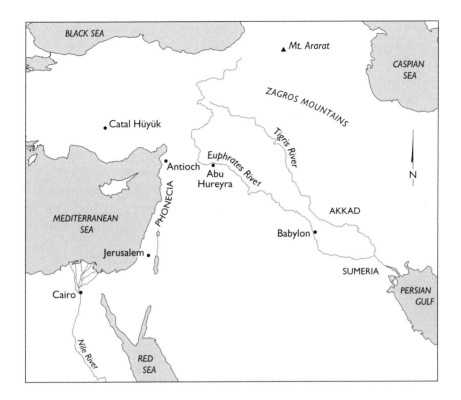

Figure 3. Map of the Middle East.

warmer climate. Fueled by growing harvests, within a couple thousand years the village's population swelled to between four thousand and six thousand.

The climate shift of the Younger Dryas was not the only factor that influenced the adoption of agriculture. Population growth during the preceding several thousand years led to the advent of sedentary communities of hunter-gatherers and contributed to the effect of this climate shift on human populations. Still, the starving people of Abu Hureyra could never have imagined that their attempt to adapt to a drying world would transform the planet.

Such adaptation may have occurred around the region. The end of the Younger Dryas coincides with changes in culture and settlement patterns throughout much of the Middle East. Neolithic settlements that emerged after the Younger Dryas were located at sites ideally suited for agriculture with rich soils and ample water supplies. Charred remains of domesticated wheat dating from 10,000 years ago are found in sites near Damascus, in northwestern Jordan, and on the Middle Euphrates River. Domesticated crops then spread south to Jericho in the Jordan Valley and northwest into southern Turkey.

Although tradition places agriculture in the Middle East long before any parallel activity in Asia and the Americas, recent research suggests that people in South America, Mexico, and China may have domesticated plants long before the first signs of settled villages in these regions. Sediments in a cave called Diaotonghuan along China's Yangtze River tell a story similar to that of Abu Hureyra in which wild rice was domesticated around the time of the Younger Dryas. Perhaps the abrupt climate changes of the Younger Dryas pushed semisettled people with declining resource bases into agricultural experimentation.

Once the climate improved, groups adapted to growing grains had an advantage. Increasing reliance on domesticated crops spread across the region. The Natufian culture that flourished along the Mediterranean coast in modern Israel, Lebanon, and Syria from 9000 until 7500 BC was based on harvesting wild grain and herding goats and gazelles. Neither plants nor animals were fully domesticated when Natufian culture arose, yet by the end of the era, hunting accounted for just a fraction of the food supply.

The regional population began to grow dramatically as domestication of wheat and legumes increased food production. By about 7000 BC small farming villages were scattered throughout the region. Communities became increasingly sedentary as intensive exploitation of small areas discouraged continuing the annual cycle of moving among hunting camps scattered around a large territory. By about 6500 BC large towns of up to several thousand people became common. The seasonal rhythm of an annual trek to follow resources was over in the Middle East.

Populations able to wrest more food from their environment could better survive periods of stress—like droughts or extreme cold. When bad times came, as they inevitably did, chance favored groups with experience tending gardens. They better endured hardships and prospered during good times. And agricultural success upped the ante. Development of more intensive and effective subsistence methods allowed human populations to grow beyond what could be supported by hunting and gathering. Eventually, communities came to depend on enhancing the productivity of natural ecosystems just to stay even, let alone grow. Early cultivators became tied to a place because mobility did not allow for tending and harvesting crops. Once humanity started down the agricultural road there was no turning back.

Learning to support more people on less land once they settled into a region, farmers could always marshal greater numbers to defeat foragers in contests over territory. As their numbers grew farmers became unbeatable on their own turf. Field by field, farms expanded to cover as much of the land as could be worked with the technology of the day.

Most farm animals were domesticated from about 10,000 to 6000 BC. My favorite exception, the dog, was brought into the human fold more than twenty thousand years earlier. I can easily imagine the scenario in which a young wolf or orphaned puppies would submit to human rule and join a pack of human hunters. Watching dogs run in Seattle's off-leash parks, I see how hunters could use dogs as partners in the hunt, especially the ones that habitually turn prey back toward the pack. In any case, dogs were not domesticated for direct consumption. There is no evidence that early people ate their first animal allies. Instead, dogs increased human hunting efficiency and probably served as sentries in early hunting camps. (Cats were relative latecomers, as they moved into agricultural settlements roughly four thousand years ago, soon after towns first overlapped with their range. As people settled their habitat, cats faced a simple choice: starve, go somewhere else, or find food in the towns. No doubt early farmers appreciated cats less for their social skills than for their ability to catch the small mammals that ate stored grain.)

Sheep were domesticated for direct consumption and economic exploitation sometime around 8000 BC, several hundred years before domestication of wheat and barley. Goats were domesticated at about the same time in the Zagros Mountains of western Iran. It is possible that seeds for the earliest of these crops were gathered to grow livestock fodder.

Cattle were first domesticated in Greece or the Balkans about 6000 B.C. They rapidly spread into the Middle East and across Europe. A revolutionary merger of farming and animal husbandry began when cattle reached the growing agricultural civilizations of Mesopotamia. With the development of the plow, cattle both worked and fertilized the fields. Conscription of animal labor increased agricultural productivity and allowed human populations to grow dramatically. Livestock provided labor that freed part of the agricultural population from fieldwork.

The contemporaneous development of crop production and animal husbandry reinforced each other; both allowed more food to be produced. Sheep and cattle turn parts of plants we can't eat into milk and meat.

Domesticated livestock not only added their labor to increase harvests, their manure helped replenish soil nutrients taken up by crops. The additional crops then fed more animals that produced more manure and led in turn to greater harvests that fed more people. Employing ox power, a single farmer could grow far more food than needed to feed a family. Invention of the plow revolutionized human civilization and transformed Earth's surface.

There were about four million people on Earth when Europe's glaciers melted. During the next five thousand years, the world's population grew by another million. Once agricultural societies developed, humanity began to double every thousand years, reaching perhaps as many as two hundred million by the time of Christ. Two thousand years later, millions of square miles of cultivated land support almost six and a half billion people—5 to 10 percent of all the people who ever lived, over a thousand times more folks than were around at the end of the last glaciation.

The new lifestyle of cultivating wheat and barley and keeping domesticated sheep spread to central Asia and the valley of the Nile River. The same system spread to Europe. Archaeological records show that between 6300 and 4800 BC adoption of agriculture spread steadily west through Turkey, into Greece, and up the Balkans at an average pace of about half a mile per year. Other than cattle, plants and animals that form the basis for European agriculture came from the Middle East.

The first farmers relied on rainfall to water their crops on upland fields. They were so successful that by about 5000 BC the human population occupied virtually the entire area of the Middle East suitable for dryland farming. The pressure to produce more food intensified because population growth kept pace with increasing food production. This, in turn, increased pressure to extract more food from the land. Not long after the first communities settled into an agricultural lifestyle, the impact of topsoil erosion and degraded soil fertility—caused by intensive agriculture and goat grazing—began to undermine crop yields. As a direct result, around 6000 BC whole villages in central Jordan were abandoned.

When upland erosion and the growing population in the Zagros Mountains pushed agricultural communities into lowlands with inadequate rainfall to grow crops, the urgent need to cultivate these increasingly marginal areas led to a major revolution in agricultural methods: irrigation. Once farmers moved into the northern portion of the floodplain between the Tigris and Euphrates rivers and began irrigating their crops, they reaped bigger harvests. Digging and maintaining canals to water their fields, set-

Figure 4. Early Mesopotamian representation of a plow from a cylinder seal (drawn from the photo of a cylinder seal rolling in Dominique Collon, *First Impressions: Cylinder Seals in the Ancient Near East* [Chicago: University of Chicago Press, 1987], 146, fig. 616).

tlements spread south along the floodplain, sandwiched between the Arabian Desert and semiarid mountains poorly suited for agriculture. As the population rose, small towns filled in the landscape, plowing and planting more of the great floodplain.

This narrow strip of exceptionally fertile land produced bumper crops. But the surpluses depended upon building, maintaining, and operating the network of canals that watered the fields. Keeping the system going required both technical expertise and considerable organizational control, spawning the inseparable twins of bureaucracy and government. By about 5000 BC people with a relatively common culture in which a religious elite oversaw food production and distribution populated nearly all of Mesopotamia—the land between two rivers.

All the good, fertile land in Mesopotamia was under cultivation by 4500 BC. There was nowhere else to expand once agriculture reached the coast. Running out of new land only intensified efforts to increase food production and keep pace with the growing population. About the time the whole floodplain came under cultivation, the plow appeared on the Sumerian plains near the Persian Gulf: it allowed greater food production from land already farmed.

Towns began to coalesce into cities. The town of Uruk (Erech) absorbed the surrounding villages and grew to about 50,000 people by 3000 BC. Construction of huge temples attests to the ability of religious leaders to marshal labor. In this initial burst of urbanization, eight major cities dominated the southern Mesopotamian region of Sumer. The population crowding into the irrigated floodplain was now a sizable proportion of humanity. Whereas hunting and gathering groups generally regarded resources as owned by and available to all, the new agricultural era permitted an unequal ownership of land and food. The first nonfarmers had appeared.

Class distinctions began to develop once everyone no longer had to work the fields in order to eat. The emergence of religious and political classes that oversaw the distribution of food and resources led to development of administrative systems to collect food from farmers and redistribute it to other segments of society. Increasing specialization following the emergence of social classes eventually led to the development of states and governments. With surplus food, a society could feed priests, soldiers, and administrators, and eventually artists, musicians, and scholars. To this day, the amount of surplus food available to nonfarmers sets the level to which other segments of society can develop.

The earliest known writing, cuneiform indentations baked into clay tablets, comes from Uruk. Dating from about 3000 BC, thousands of such tablets refer to agricultural matters and food allocation; many deal with food rationing. Writing helped a diversifying society manage food production and distribution, as population kept pace with food production right from the start of the agricultural era.

Rivalries between cities grew along with their populations. The organization of militias reflects the concentration of wealth that militarized Mesopotamian society. Huge walls with defensive towers sprang up around cities. A six-mile-long wall circling Uruk spread fifteen feet thick. Wars between Sumerian city-states gave rise to secular military rulers who crowned themselves as the governing authority. As the new rulers appropriated land from the temples and large estates became concentrated in the hands of influential families and hereditary rulers, the concept of private property was born.

The few million acres of land between the Tigris and Euphrates rivers fed a succession of civilizations as the rich valley turned one conquering horde after another into farmers. Empires changed hands time and again, but unlike soils on the mountain slopes where agriculture began, the rich floodplain soil did not wash away when cleared and planted. Coalescence of Sumerian cities into the Babylonian Empire about 1800 BC represented the pinnacle of Mesopotamian organizational development and power. This merger solidified a hierarchical civilization with formalized distinctions recognizing legal classes of nobility, priests, peasants, and slaves.

But the irrigation that nourished Mesopotamian fields carried a hidden risk. Groundwater in semiarid regions usually contains a lot of dissolved salt. Where the water table is near the ground surface, as it is in river valleys and deltas, capillary action moves groundwater up into the soil to evaporate, leaving the salt behind in the ground. When evaporation rates are high, sustained irrigation can generate enough salt to eventually poison crops. While irrigation dramatically increases agricultural output, turning sun-baked floodplains into lush fields can sacrifice long-term crop yields for short-term harvests.

Preventing the buildup of salt in semiarid soils requires either irrigating in moderation, or periodically leaving fields fallow. In Mesopotamia, centuries of high productivity from irrigated land led to increased population density that fueled demand for more intensive irrigation. Eventually, enough salt crystallized in the soil that further increases in agricultural production were not enough to feed the growing population.

The key problem for Sumerian agriculture was that the timing of river runoff did not coincide with the growing season for crops. Flow in the Tigris and Euphrates peaked in the spring when the rivers filled with snow melt from the mountains to the north. Discharge was lowest in the late summer and early fall when new crops needed water the most. Intensive agriculture required storing water through soaring summer temperatures. A lot of the water applied to the fields simply evaporated, pushing that much more salt into the soil.

Salinization was not the only hazard facing early agricultural societies. Keeping the irrigation ditches from silting up became a chief concern as extensive erosion from upland farming in the Armenian hills poured dirt into the Tigris and Euphrates. Conquered peoples like the Israelites were put to work pulling mud from the all-important ditches. Sacked and rebuilt repeatedly, Babylon was finally abandoned only when its fields became too difficult to water. Thousands of years later piles of silt more than thirty feet high still line ancient irrigation ditches. On average, silt pouring out of the rivers into the Persian Gulf has created over a hundred feet of new land a year since Sumerian time. Once a thriving seaport, the ruins of Abraham's hometown of Ur now stand a hundred and fifty miles inland.

As Sumer prospered, fields lay fallow for shorter periods due to the growing demand for food. By one estimate almost two-thirds of the thirty-five thousand square miles of arable land in Mesopotamia were irrigated when the population peaked at around twenty million. The combination of a high load of dissolved salt in irrigation water, high temperatures during the irrigation season, and increasingly intensive cultivation pumped ever more salt into the soil.

Temple records from the Sumerian city-states inadvertently recorded agricultural deterioration as salt gradually poisoned the ground. Wheat, one of the major Sumerian crops, is quite sensitive to the concentration of salt in the soil. The earliest harvest records, dating from about 3000 BC, report equal amounts of wheat and barley in the region. Over time the proportion of wheat recorded in Sumerian harvests fell and the proportion of barley rose. Around 2500 BC wheat accounted for less than a fifth of the harvest. After another five hundred years wheat no longer grew in southern Mesopotamia.

Wheat production ended not long after all the region's arable land came under production. Previously, Sumerians irrigated new land to offset shrinking harvests from salty fields. Once there was no new land to cultivate, Sumerian crop yields fell precipitously because increasing salinization meant that each year fewer crops could be grown on the shrinking amount of land that remained in production. By 2000 BC crop yields were down by half. Clay tablets tell of the earth turning white in places as the rising layer of salt reached the surface.

The decline of Sumerian civilization tracked the steady erosion of its agriculture. Falling crop yields made it difficult to feed the army and maintain the bureaucracy that allocated surplus food. As their armies deteriorated, the independent city-states were assimilated by the younger Akkadian empire from northern Mesopotamia at the time of the first serious decline in crop yields around 2300 BC. During the next five hundred years the region fell to a succession of conquerors. By 1800 BC crop yields were down to a third of the initial yields and southern Mesopotamia declined into an impoverished backwater of the Babylonian Empire. Salinization that destroyed the Sumerian city-states spread northward, triggering an agricultural collapse in central Mesopotamia between 1300 and 900 BC.

Mesopotamian agricultural practices also spread west into North Africa along the Mediterranean coast and into Egypt. The valley of the Nile provides a notable exception to the generality that civilizations prosper for only a few dozen generations. The first farming settlements in the Nile delta date from about 5000 BC. Farming and livestock herding gradually replaced hunting and gathering as silt carried by the river began building a broad, seasonally flooded, and exceptionally fertile delta once the postglacial sea level's rise slowed enough to let the silt pile up in one place. At

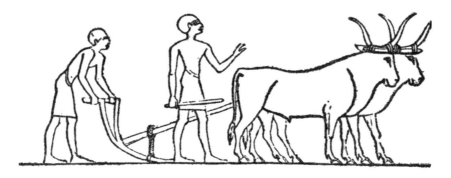

Figure 5. Ancient Egyptian plow (Whitney 1925).

first Egyptian farmers simply cast seeds into the mud as the annual flood receded, harvesting twice the amount of grain used for seed. Thousands of people died when the water drained too quickly and crops failed. So farmers started impounding water behind dikes, forcing it to sink into the rich earth. As the population grew, innovations like canals and water wheels irrigated land higher and farther from the river, allowing more people to be fed.

The floodplain of the Nile proved ideal for sustained agriculture. In contrast to Sumerian agriculture's vulnerability to salinization, Egyptian agriculture fed a succession of civilizations for seven thousand years, from the ancient pharaohs through the Roman Empire and into the Arab era. The difference was that the Nile's life-giving flood reliably brought little salt and a lot of fresh silt to fields along the river each year.

The geography of the river's two great tributaries mixed up the ideal formula to nourish crops. Each year the Blue Nile brought a twentieth of an inch (about a millimeter) of silt eroded from the Abyssinian highlands. The White Nile brought humus from central Africa's swampy jungles. Fresh silt replaced mineral nutrients used by the previous crop and the influx of humus refreshed soil organic matter that decayed rapidly under the desert sun. In addition, the heavy rains that fell during June in the uplands to the south produced a flood that reliably reached the lower Nile in September and subsided in November, just the right time for planting crops. The combination produced abundant harvests year after year.

Egyptian irrigation exploited a natural process through which overflow channels spread floodwaters across the valley. Irrigating fields did not require elaborate canals; instead the river's natural levees were breached to direct water to particular places on the floodplain. After the annual flood, the water table dropped more than ten feet below the valley bottom, eliminating the threat of salinization. In contrast to the experience of Mesopotamian farmers, Egyptian wheat harvests increased over time. The longevity of Egyptian agriculture reflects a system that took advantage of the natural flood regime with minimal modification.

Fresh dirt delivered by predictable annual floods meant that fields could be kept in continuous production without compromising soil fertility. But the population was still subject to the whims of the climate. A few bad years, or even a single disastrous one, could be catastrophic. Extended drought severely reduced crop yields; a peasant revolt during one from about 2250 to 1950 BC toppled the Old Kingdom. Still, the generally reliable Nile sustained a remarkably successful agricultural endeavor.

Unlike in Mesopotamia, regulating the distribution of the river's annual floodwaters remained a local responsibility. There was little impetus for developing a centralized authority. Class distinctions and division of labor developed in Egypt only after the adoption of perennial irrigation to produce cash crops had undermined traditional village communities. The despotic political superstructure of Mesopotamia was not an inevitable result of hydraulic civilization.

Eventually, however, the agricultural surplus fueled the growth of an administrative and political elite. Egypt coalesced into a unified state about 3000 BC, developing into an ancient superpower that rivaled Mesopotamia. The rise of commercial farming not only allowed the population to grow, it meant that they had to be kept occupied. Some even suggest that the Great Pyramids were public works projects intended to combat unemployment.

Egyptian agriculture remained remarkably productive for thousands of years until people adopted new approaches out of tune with the river's natural rhythm. Desire to grow cotton for export to Europe brought aggressive year-round irrigation to the Nile in the early nineteenth century. Just as in the scenario that unfolded thousands of years earlier in Mesopotamia, salt began to build up in the soil as the water table rose below overly irrigated fields. By the 1880s British agricultural expert Mackenzie Wallace described irrigated fields covered by white salts "covering the soil and glistening in the sun like untrodden snow."[1] As dramatic as this spectacle appeared, the adverse effects of irrigation were dwarfed by those of damming the Nile.

In the past half century, civilization finally acquired the engineering skill to cripple an almost indestructible land. After four years of work, Egyptian president Gamal Abdel Nasser and Soviet premier Nikita Khrushchev watched Soviet engineers divert the Nile in May 1964 to build the Aswan High Dam. Two and a half miles across, and more than seventeen times as massive as the Great Pyramid, the dam impounds a lake 300 miles long and 35 miles wide that can hold twice the river's annual flow.

The British hydrologists who controlled Egypt's river until the 1952 coup that brought Nasser to power opposed building the dam because evaporation would send too much of the huge new lake back into the sky. Their fears were well founded. Under the desert sun six feet of water evaporates off the top of the lake each year—more than fourteen cubic kilometers of water that used to head down the river. But a greater problem was that the 130 million tons of dirt that the Nile carried off from Ethiopia settled out at the bottom of Lake Nasser.

After advancing for thousands of years since sea level stabilized, the Nile delta is now eroding, cut off from a supply of silt. Although the dam allows farmers to grow two or three crops a year using artificial irrigation, the water now delivers salt instead of silt. A decade ago salinization had already reduced crop yields from a tenth of the fields on the Nile delta. Taming the Nile disrupted the most stable agricultural environment on Earth.

As the renowned fertility of the Nile valley began to fall, agricultural output was sustained with chemical fertilizers that peasant farmers could not afford. Modern farmers along the Nile are some of the world's foremost users of chemical fertilizers—conveniently produced in new factories that are among the largest users of power generated by Nasser's dam. Now, for the first time in seven thousand years, Egypt—home of humanity's most durable garden—imports most of its food. Still, the remarkable longevity of Egyptian civilization is a primary exception to the general rise-and-fall of ancient civilizations.

The history of Chinese agriculture provides another example where, as in Mesopotamia, dryland farmers from the uplands moved down onto floodplains as the population exploded. Unlike the Sumerians who appear to have treated all soils the same, the Yao dynasty (2357–2261 BC) based taxation on a soil survey that recognized nine distinct types of dirt. A later soil classification, dating from 500 BC, codified older ideas based on soil color, texture, moisture, and fertility.

Today, the Chinese people overwhelmingly live on the alluvial plains where great rivers descending from the Tibetan Plateau deposit much of their load of silt. Flooding has been a problem for thousands of years on the Huanghe, better known in the West as the Yellow River, a name imparted by the color of dirt eroded from the river's deforested headwaters. Before the first levees and dikes were constructed in 340 BC, the river meandered across a broad floodplain. In the second century BC the river's Chinese name changed from Great River to Yellow River when the sediment load increased tenfold as farmers began plowing up the highly erodible silty (loess) soils into the river's headwaters.

The earliest communities along the Yellow River were situated on elevated terraces along tributaries. Only later, after the area became densely populated, did people crowd onto the floodplain. Extensive levees to protect farmlands and towns along the river kept floodwaters, and the sediment they carried, confined between the levees. Where the river hit the plains, the weakening current began dropping sediment out between the levees instead of across the floodplain. Rebuilding levees ever higher to contain the floodwaters ensured that the riverbed climbed above the alluvial plain about a foot every century.

By the 1920s the surface of the river towered thirty feet above the floodplain during the high-water season. This guaranteed that any flood that breached the levees was devastating. Floodwaters released from the confines of the levees roared down onto the floodplain, submerging farms, towns, and sometimes even whole cities beneath a temporary lake. In 1852 the river jumped its dikes and flowed north, flooding cities and villages and killing millions of people before draining out hundreds of miles to the north. More than two million people drowned or died in the resulting famine when the river breached its southern dike and submerged the province of Henan during the flood of 1887–89. With the river flowing high above its floodplain, levee breaches are always catastrophic.

Soil erosion in northern China grabbed international attention when a withering drought killed half a million people in 1920–21. Some twenty million people were reduced to eating literally anything that grew from the soil. In some areas starving people stripped the landscape down to bare dirt. The ensuing erosion triggered mass migrations when fields blew away. But this was not unusual. A 1920s famine-relief study documented that famine had occurred in some part of China during each of the previous two thousand years.

In 1922 forester and Rhodes scholar Walter Lowdermilk took a job at the University of Nanking to work on famine prevention in China. Touring the country, he deduced how soil abuse had influenced Chinese society. The experience impressed upon him the fact that soil erosion could cripple civilizations. Years later, after traveling widely to study soil erosion in Asia, the Middle East, and Europe, Lowdermilk described his profession as reading "the record which farmers, nations, and civilizations have written in the land."[2]

Approaching the site where the Yellow River broke through its dikes in 1852, Lowdermilk described how a huge flat-topped hill rose fifty feet above the alluvial plain to dominate the horizon. Climbing up to this elevated plain inside the river's outer levee, Lowdermilk's party traversed seven miles of raised land before coming to the inner dike and then the river itself. Over thousands of years, millions of farmers armed with baskets full of dirt walled in and gradually raised four hundred miles of the river above its floodplain and delta. Seeing the muddy yellowish water, Lowdermilk realized that the heavy load of silt eroded from the highlands began to settle out when the river's slope dropped to less than one foot per mile. The more silt built up the riverbed, the faster farmers raised the dikes. There was no winning this game.

Determined to find the source of the dirt filling in the river, Lowdermilk traveled upstream to the province of Shansi (Shanxi), the cradle of Chinese civilization. There in northwestern China he found a landscape deeply incised by gullies, where intensive cultivation after clearing of forests from steep, highly erodible slopes was sending the soil downstream. Lowdermilk was convinced that deforestation alone would not cause catastrophic erosion—shrubs and then trees simply grew back too fast. Instead, farmers cultivating steep slopes left the soil vulnerable to erosion during intense summer downpours. "Erosion is only indirectly related to the destruction of the former extensive forests, but is directly related to the cultivation of the slope lands for the production of food crops."

Rather than the axe, the plow had shaped the region's fate, as Lowdermilk observed. "Man has no control over topography and little over the type of rainfall which descends on the land. He can, however, control the soil layer, and can, in mountainous areas, determine quite definitely what will become of it."[3] Lowdermilk surmised how the early inhabitants of the province cleared the forest from easily tilled valley bottoms. Farms spread higher up the slopes as the population grew; Lowdermilk even found evidence for abandoned fields on the summits of high mountains. Viewing the effects of farming the region's steep slopes, he concluded that summer rains could strip fertile soil from bare, plowed slopes in just a decade or two. Finding abundant evidence for abandoned farms on slopes throughout the region, he concluded that the whole region had been cultivated at some time in the past. The contrast of a sparse population and extensive abandoned irrigation systems told of better days gone by.

Lowdermilk had first recognized the impact of people on the lands of northern China at a virtually abandoned walled city in the upper Fen River valley. Surveying the surrounding land, he discerned how the first inhabitants occupied a forested landscape blanketed by fertile soil. As the population prospered and the town grew into a city, the forest was cleared and fields spread from the fertile valley bottoms up the steep valley walls. Topsoil ran off the newly cleared farms pushing up the mountain sides. Eventually, goats and sheep grazing on the abandoned fields stripped the remaining soil from the slopes. Soil erosion so undercut agricultural productivity that the people either starved or moved, abandoning the city.

Lowdermilk estimated than a foot of topsoil had been lost from hundreds of millions of acres of northern China. He found exceptions where Buddhist temples protected forests from clearing and cultivation; there the exceptionally fertile forest soil was deep black, rich in humus. Lowdermilk described how farmers were clearing the remaining unprotected forest to farm this rich dirt, breaking up sloping ground with mattocks to disrupt tree roots and allow plowing. At first, plowing smoothed over new rills and gullies, but every few years erosion pushed farmers farther into the forest in search of fresh soil. Seeing how colonizing herbs and shrubs shielded the ground as soon as fields were abandoned, Lowdermilk blamed the loss of the soil on intensive plowing followed by overgrazing. He concluded that the region's inhabitants were responsible for impoverishing themselves—just too slowly for them to notice.

Over the next three years, Lowdermilk measured erosion rates from protected groves of trees, on farm fields, and from fields abandoned because of erosion. He found that runoff and soil erosion on cultivated fields were many times greater than under the native forest. Farmers in the headwaters of the Yellow River were increasing the river's naturally high sediment load, exacerbating flooding problems for people living downstream.

Today the cradle of Chinese civilization is an impoverished backwater lacking fertile topsoil, just like Mesopotamia and the Zagros Mountains. Both of these ancient civilizations started off farming slopes that lost soil, and then blossomed when agriculture spread downstream to floodplains that could produce abundant food if cultivated.

Another commonality among agricultural societies is that the majority of the population lives harvest-to-harvest with little to no hedge against crop failure. Throughout history, our growing numbers kept pace with agricultural production. Good harvests tended to set population size, making a squeeze inevitable during bad years. Until relatively recently in the agricultural age, this combination kept whole societies on the verge of starvation.

For over 99 percent of the last two million years, our ancestors lived off the land in small, mobile groups. While certain foods were likely to be in short supply at times, it appears that some food was available virtually all the time. Typically, hunting and gathering societies considered food to belong to all, readily shared what they had, and did not store or hoard—egalitarian behavior indicating that shortages were rare. If more food was needed, more was found. There was plenty of time to look. Anthropologists generally contend that most hunting and gathering societies had relatively large amounts of leisure time, a problem few of us are plagued with today.

Farming's limitation to floodplains established an annual rhythm to early agricultural civilizations. A poor harvest meant death for many and hunger for most. Though most of us in developed countries are no longer as directly dependent on good weather, we are still vulnerable to the slowly accumulating effects of soil degradation that set the stage for the decline of once-great societies as populations grew to exceed the productive capacity of floodplains and agriculture spread to the surrounding slopes, initiating cycles of soil mining that undermined civilization after civilization.

FOUR

Graveyard of Empires

To Protect Your Rivers, Protect Your Mountains

EMPEROR YU (CHINA)

IN THE EARLY 1840S NEW YORK LAWYER, adventurer, and amateur archaeologist John Lloyd Stephens found the ruins of more than forty ancient cities in dense Central American jungle. After excavating at Copán in Guatemala, traveling north to Mexico's ruined city of Pelenque, and returning to the Yucatán, Stephens realized that the jungle hid a lost civilization. His revelation shocked the American public. Native American civilizations rivaling those of the Middle East didn't fit into the American vision of civilizing a primeval continent.

A century and a half after Stephens's discovery, I stood atop the Great Pyramid at Tikal and relived his realization that the surrounding hills were ancient buildings. The topography itself outlined a lost city, reclaimed by huge trees, roots locked around piles of hieroglyphic-covered rubble. Temple-top islands rising above the forest canopy were the only sign of an ancient tropical empire.

With different characters and contexts, Tikal's story has been repeated many times around the world—in the Middle East, Europe, and Asia. The capital of many a dead civilization lives off tourism. Did soil degradation destroy these early civilizations? Not directly. But time and again it left societies increasingly vulnerable to hostile neighbors, internal sociopolitical disruption, and harsh winters or droughts.

Although societies dating back to ancient Mesopotamia damaged their environments, dreams of returning to a lost ethic of land stewardship still underpin modern environmental rhetoric. Indeed, the idea that ancient peoples lived in harmony with the environment remains deeply rooted in the mythology of Western civilizations, enshrined in the biblical imagery of the garden of Eden and notions of a golden age of ancient Greece. Yet few societies managed to conserve their soil—whether deliberately or through traditions that defined how people treated their land while farms filled in the landscape and villages coalesced into towns and cities. With allowances for different geographical and historical circumstances, the story of many civilizations follows a pattern of slow, steady population growth followed by comparatively abrupt societal decline.

Ancient Greece provides a classic example of too much faith in stories of lost utopias. Hesiod, a contemporary of Homer, wrote the earliest surviving description of Greek agriculture about eight centuries before the time of Christ. Even the largest Greek estates produced little more than needed to feed the master, his slaves, and their respective families. Like Ulysses' father, Laertes, the early leaders of ancient Greece worked in their own fields.

Later, in the fourth century BC, Xenophon wrote a more extensive description of Greek agriculture. By then wealthy landowners employed superintendents to oversee laborers. Still, Xenophon advised proprietors to observe what their land could bear. "Before we commence the cultivation of the soil, we should notice what crops flourish best upon it; and we may even learn from the weeds it produces what it will best support."[1] Xenophon advised farmers to enrich their soil both with manure and with burned crop stubble plowed back into the fields.

Ancient Greeks knew about the fertilizing properties of manure and compost, but it is not clear how widely such practices were followed. Even so, for centuries after the revival of classical ideals during the European Renaissance, historians glorified the ancient Greeks as careful stewards of their land. But the dirt of modern Greece tells a different story—a tale of destructive episodes of soil erosion.

With thin rocky soils covering much of its uplands, only about a fifth of Greece could ever support agriculture. The adverse effects of soil erosion on society were known in classical times; the Greeks replenished soil nutrients and terraced hillside fields to retard erosion. Nonetheless, the hills around Athens were stripped bare by 590 BC, motivating concern over how to feed the city. Soil loss was so severe that Solon, the famed reformer of the constitution, proposed a ban on plowing steep slopes. By the time of the Peloponnesian War (431–404 BC), Egypt and Sicily grew between a third and three-quarters of the food for Greek cities.

Plato (427–347 BC) attributed the rocky slopes of his native Attica to pre-Hellenistic soil erosion following deforestation. He also commented on soil's key role in shaping Athenian society, maintaining that the soils of earlier times were far more fertile. Plato held that the soil around Athens was but a shadow of its former self, citing evidence that bare slopes were once forested. "The rich, soft soil has all run away leaving the land nothing but skin and bone. But in those days the damage had not taken place, the hills had high crests, the rocky plain of Phelleus was covered with rich soil, and the mountains were covered by thick woods, of which there are some traces today."[2] Seeing how harvesting the natural fertility of the surrounding land allowed Athens to blossom into a regional power, Plato held that the root of his city's wealth lay in its soil.

Aristotle (384–322 BC) shared Plato's conviction that Bronze Age land use degraded soil productivity. His student Theophrastus (371–286 BC) recognized six distinct types of soil composed of different layers, including a humus-rich layer above subsoil that supplied nutrients to plants. Theophrastus made a point of distinguishing fertile topsoil from the underlying earth.

Both Plato and Aristotle recognized signs that Bronze Age land use had degraded their region's soil. Several thousand years and several civilizations later, archaeologists, geologists, and paleoecologists vindicated Aristotle's estimate of the timing: farmers arrived about 5000 BC and dozens of agricultural settlements were scattered throughout the region by 3000 BC; cultivation intensified about the time Aristotle posited the first serious effects of soil erosion there. Such knowledge, however, did not prevent classical Greece from repeating the pattern.

Over the past several decades, studies of soils throughout Greece—from the Argive Plain and the southern Argolid in the Peloponnese to Thessaly and eastern Macedonia—showed that even the dramatic climate change at the end of the last glaciation did not increase erosion. Instead, thick forest soils developed in the warming climate as oak forest replaced grassland across the Greek countryside. Over thousands of years the soil grew half a foot to several feet thick depending on local conditions. Soil erosion began to exceed soil production only after introduction of the plow.

The first Greek settlements were located in valleys with good soils near reliable water supplies. As the landscape filled with people, farmers began

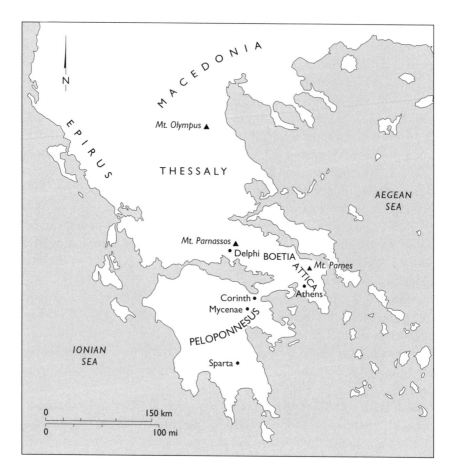

Figure 6. Map of ancient Greece.

advancing onto steeper, less productive slopes. Extensive tilling and grazing stripped soil from hillsides and piled thick deposits of reworked dirt in valleys. Ancient agricultural artifacts can still be found on the rocky slopes of areas that lack enough soil to grow much vegetation.

Sediments trapped in valley bottoms, and remnant pockets of soil on the slopes themselves, record cycles of erosion and soil formation throughout Greece. The deepest layers of valley-filling sediments date from glacial to interglacial climate changes during the past quarter million years. Higher layers in the stack of dirt tell of more recent episodes of hillslope erosion as well as intervening periods when soils developed. The first postglacial

Figure 7. Parthenon. Albumen print by William James Stillman, 1869 (courtesy of Research Library, the Getty Research Institute, Los Angeles, California [92.R.84]).

deposits of reworked hillslope soils in the valleys generally date from the Bronze Age arrival of agriculture. Erosional episodes similar in outline, but different in detail, occurred across ancient Greece where farming spread out of the valleys and onto hillslopes.

Soils of the southern Argolid, for example, record four periods of post-glacial erosion during times of intensive land use. The first, from roughly 4500 to 3500 BC, was a time when thick woodland soils were widely settled by early farmers. Introduction of the plow and the spread of farming into steeper terrain led to widespread erosion around 2300 to 1600 BC. Hillslope soils gradually rebuilt during the dark age before the rise of classical Greek civilization. The area was again densely settled in late Roman times and another period of depopulation followed in the seventh century AD. About fifteen inches of soil are estimated to have been lost from Argolid uplands since the start of Bronze Age agriculture. As many as three feet of soil may have been stripped from some lowland slopes.

Valley bottom sediments of the Argive Plain in the northeastern Peloponnese also testify to four periods of extensive soil erosion in the past five thousand years. Today, thick red and brown soils are found only in hollows and at the foot of slopes protected from streams. Remnants of hillslope soils and archaeological evidence show that since the Bronze Age there have been centuries-long periods with high settlement density, intensive farming, and accelerated soil erosion separated by millennia-long periods of low population density and soil formation.

Alexander the Great's homeland of Macedonia in eastern Greece underwent similar episodes of soil erosion accompanied by stream filling, and followed by landscape stability. The pace of soil erosion doubled in the late Bronze Age, and then doubled again from the third century BC to the seventh century AD. Another round started after the fifteenth century—defining a cycle with a roughly thousand-year periodicity, just as in other parts of Greece.

Regional climate changes cannot explain the boom-and-bust pattern of human occupation in ancient Greece because the timing of land settlement and soil erosion differed around the region. Instead, modern geoarchaeological surveys show that soil erosion episodically disrupted local cultures, forced settlements to relocate, led to changes in agricultural practices, and caused periodic abandonment of entire areas.

An ancient geopolitical curiosity provides further evidence that people destroyed Greek soils. The northern slopes of Mount Parness define the border between Boeotia and Attica. Oddly, the region belonged to Attica but was accessible only from Boeotia. So the region remained forested because Athenians could not get to it and Boeotians could not use it. While both city-states suffered severe soil erosion in their cultivated heartland, the no-man's-land on the border still retains a thick forest soil.

Extensive Bronze Age soil erosion coincides with changing agricultural practices that allowed a major increase in human population. The transition from highly localized, spring-fed agriculture using digging sticks to rain-fed agriculture based on clearing and plowing whole landscapes fueled an expansion of settlements. Initially, very low hillslope erosion rates increased slowly as agriculture spread until eventually erosion increased tenfold during the Bronze Age. Subsequently, erosion rates dropped back to close to the natural rate before once again increasing tenfold during the classical and Roman eras.

Almost the entire landscape was cultivated by classical times. Massive piles of dirt deposited in valley bottoms document extensive erosion of forest soils from hillsides disturbed by initial agricultural colonization. In places, later episodes of soil erosion were not as severe because continued farming and grazing prevented rebuilding thick soils. Even so, ancient erosion control measures like terraced hillsides and check dams built to slow the growth of gullies provide direct evidence of a fight to save soil.

The variety of crops excavated from Neolithic sites in Greece indicate that pre-Bronze Age agriculture was highly diversified. Sheep, goats, cows, and pigs were kept on small, intensively worked mixed-crop farms. Evidence of plow-based agriculture on estates worked by oxen records a progressive shift from diversified small-scale farms to large plantations. By the late Bronze Age, large areas controlled by palaces specialized in growing cereals. Olives and grapes became increasingly important as small farms spread into progressively more marginal areas prone to soil erosion. This was no coincidence—they grew well in thin, rocky soils.

Hesiod, Homer, and Xenophon all described two-field systems with alternate fallow years. It was normal to plow both fallow and planted fields three times a year, once in the spring, once in summer, and again in the autumn right before sowing. All this plowing gradually pushed soil downhill and left fields bare and vulnerable to erosion. Whereas Hesiod recommended using an experienced plowman who could plow a straight line regardless of the lay of the land, by later classical times terraces were built to try and retain soil and extend the productive life of hillside fields.

Modern examples show just how rapidly Greek soils can erode. On some overgrazed slopes, thickets of fifty-year-old oak standing on one-and-a-half-foot-high soil pedestals testify to modern erosion rates of just over a quarter of an inch per year. Live trees with roots exposed up to two and a half feet above the present ground surface record decades of soil erosion at around half an inch per year. When exposed to the direct effects of rainfall, land can fall apart at a rate apparent to even casual observers.

Little more than six centuries after the first Olympics were held in 776 BC, the Romans captured and destroyed Corinth, assimilating Greece into the Roman Empire in 146 BC. By then, after the second round of widespread soil erosion, Greece was no longer a major power. Some remarkable geologic detective work has shown how, like the ancient Greeks, the Romans also accelerated soil erosion enough to impact their society.

In the mid-1960s Cambridge University graduate student Claudio Vita-Finzi picked Roman potsherds from the banks of a Libyan wadi out of deposits previously thought to date from glacial times. Puzzled by the large amount of sediment so recently deposited by the stream, he poked around ancient dams, cisterns, and ruined cities and found evidence for substantial historical soil erosion and floodplain deposition. Intrigued, he set about trying to determine whether these geologic changes in historical times told of climate change or land abuse.

Traveling from Morocco north to Spain, and then back east across North Africa to Jordan, Vita-Finzi found evidence for two periods of extensive hillslope erosion and valley bottom sedimentation in river valleys around the Mediterranean. Deposits he called the Older Fill recorded erosion during late glacial times. Convinced that what he at first thought to be a Libyan curiosity was instead part of a broader pattern, Vita-Finzi attributed the younger valley fill to lower stream discharge caused by increasing aridity at the beginning of the late Roman era.

As often happens with new theories, people trying to fit additional observations into a simple framework found a more complicated story. The timing of soil erosion and valley filling differed around the region. How could Vita-Finzi's proposed regional drying affect neighboring areas at different times, let alone cause repeated episodes of erosion in some places? Just as in Greece, evidence now shows that people accelerated soil erosion in the Roman heartland as well as Roman provinces in North Africa and the Middle East. Even so, a simple choice of causes between climate and people is misleading. Droughts and intense storms accelerated erosion periodically on land where agricultural practices left soils bare and vulnerable.

As in other Paleolithic hunting cultures in southern Europe, an almost exclusive reliance on hunting large animals in central Italy gave way to more mixed hunting, fishing, and gathering as forests returned after the glaciers retreated. Thousands of years later, sometime between 5000 and 4000 BC, immigrants from the east introduced agriculture to the Italian Peninsula. Sheep, goat, and pig bones found along with wheat, barley seeds, and grinding stones reveal that these first farmers relied on mixed cereal cultivation and animal husbandry. Occupying ridges mantled with easily worked, well-drained soils these farmers relied on an integrated system of cereal cultivation and grazing similar to traditional peasant agriculture described by Roman agronomists thousands of years later. Between 3000 and 1000 BC, agricultural settlements spread across the Italian landscape.

From the early Neolithic to the end of the Bronze Age, Italian agriculture expanded from a core of prime farmlands into progressively more marginal land. The basic system of small-scale farms practicing mixed animal husbandry and growing a diversity of crops remained remarkably stable during this period of agricultural expansion—Bronze Age farmers still

Figure 8. Map of Roman Italy.

followed the practices of their Neolithic ancestors. Between about 4000 and 1000 BC, agriculture spread from the best sites used by the first farmers to steeper slopes and hard-to-work valley bottom clays.

Iron came into widespread use about 500 BC. Before then only the wealthy and the military had access to metal tools. More abundant and cheaper than bronze, iron was hard, durable, and readily formed to fit over wood. Farmers began fitting plows and spades with iron blades to carve through topsoil and down into denser subsoil. Most of Italy remained forested around 300 BC, but new metal tools allowed extensive deforestation over the next several centuries.

When Romulus founded Rome in about 750 BC, he divided up the new state into two-acre parcels, a size his followers could cultivate themselves. The soils of central Italy were famously productive when the Roman

Republic was founded in 508 BC. The average farm still consisted of roughly one to five acres (half a hectare to two hectares) of land, just enough land to feed a family. Many prominent Roman family names were derived from vegetables their ancestors excelled at growing. Calling a man a good farmer was high praise in the republic. Cincinnatus was plowing his fields when summoned to become dictator in 458 BC.

Early Roman farms were intensively worked operations where diversified fields were hoed and weeded manually and carefully manured. The earliest Roman farmers planted a multistory canopy of olives, grapes, cereals, and fodder crops referred to as *cultura promiscua*. Interplanting of understory and overstory crops smothered weeds, saved labor, and prevented erosion by shielding the ground all year. Roots of each crop reached to different depths and did not compete with each other. Instead, the multicrop system raised soil temperatures and extended the growing season. In the early republic, a Roman family could feed itself working the typical plot of land by hand. (And such labor-intensive farming is best practiced on a small scale.) Using an ox and plow saved labor but required twice as much land to feed a family. As plowing became standard practice, the demand for land increased faster than the population.

So did erosion. Extensive deforestation and plowing of the Campagna increased hillslope erosion to the point that antierosion channels were built to stabilize hillside farms. Despite such efforts, sediment-choked rivers turned valley bottoms into waterlogged marshes as plows advanced up the surrounding slopes. Malaria became a serious concern about 200 BC when silt eroded from cultivated uplands clogged the Tiber River and the agricultural valley that centuries before supported more than a dozen towns became the infamous Pontine Marshes. Large areas of worn-out hills and newly marshy valleys meant that formerly cultivated regions were becoming pastures of little use beyond grazing. Once-flourishing towns emptied as pastures supported fewer farmers than did the former fields.

Romans recognized that their wealth came from the earth; after all they coined the name Mother Earth *(mater terra)*. As did the Greeks before them, Roman philosophers recognized the fundamental problems of soil erosion and loss of soil fertility. But unlike Aristotle and Plato, who simply described evidence for past erosion, Roman philosophers exuded confidence that human ingenuity would solve any problems. Cicero crisply summarized the goal of Roman agriculture as to create "a second world within the world of nature." Yet even as Roman farmers used deeper plows and adapted their choice of crops to their denuded slopes, keeping soil on the Roman heartland became increasingly problematic. As Rome grew, Roman agriculture kept up by expanding into new territory.

Central Italy has four main types of soil: clay-rich soils prone to erosion when cultivated; limestone soils including ancient, deeply weathered Terra Rossa; fertile, well-drained volcanic soils; and valley bottom alluvial soil. Agricultural practices induced severe erosion on both clay-rich and limestone soils that mantled upland areas. The original forest soils have been so eroded in places that farmers now plow barely weathered rock. In many upland areas, limestone soils have been reduced to small residual pockets. Across much of central Italy, centuries of farming and grazing left a legacy of thin soils on bare slopes.

Roman farmers distinguished soils based on their texture (sand or clay content), structure (whether the particles grouped together as crumbs or clods), and capacity to absorb moisture. They assessed a soil's quality according to the natural vegetation that grew on it, or its color, taste, and smell. Different soils were rich or poor, free or stiff, and wet or dry. The best soil was a rich blackish color, absorbed water readily, and crumbled when dry. Good dirt did not rust plows or attract crows after plowing; if left fallow, healthy turf rapidly covered it. Like Xenophon, Roman agriculturalists understood that different things grew best in different soils; grapevines liked sandy soil, olive trees grew well on rocky ground.

Marcus Porcius Cato (234–149 BC) wrote *De agri cultura,* the oldest surviving Roman work on agriculture. Cato focused on grape, olive, and fruit growing and distinguished nine types of arable soils, subdivided into twenty-one minor classes based mainly on what grew best in them. He called farmers the ideal citizens and considered the agricultural might of its North African rival, Carthage, a direct threat to Roman interests. Carthage was an agricultural powerhouse capable of becoming a military rival. In perhaps the earliest known political stunt, Cato brought plump figs grown in Carthage onto the Senate floor to emphasize his view that "Carthage must be destroyed." Ending all his speeches, no matter what the subject, with this slogan, Cato's agitating helped trigger the Third Punic War (149–146 BC) in which Carthage was torched, her inhabitants slaughtered, and her fields put to work feeding Rome.

Cato's businesslike approach to farming appears tailored to help Rome's rising class of plantation owners maximize wine and olive oil harvests while keeping costs to a minimum. Low-tech versions of the plantation agriculture of colonial and modern times, the agrarian enterprises he described became specialized operations with a high degree of capital investment.

Falling slave and grain prices began driving tenant farmers off the land and encouraged raising cash crops on large estates using slave labor.

The next surviving Roman agricultural text dates from about a century later. Born on a farm in the heart of rural Italy, Marcus Terentius Varro (116–27 BC) wrote *De re rustica* at a time when these large estates dominated the Roman heartland. Varro himself owned an estate on the slopes of Vesuvius. Recognizing almost one hundred types of soil, he advocated adapting farming practices and equipment to the land. "It is also a science, which explains what crops are to be sown and what cultivations are to be carried out in each kind of soil, in order that the land may always render the highest yields."[3] Like most Roman agricultural writers, Varro emphasized obtaining the highest possible yields through intensive agriculture.

Although cereals grew best in the alluvial plains, Italy's lowland forest was already cleared and cultivated by Varro's time. Increasing population had pushed cereal cultivation into the uplands as well. Varro noted that Roman farmers grew cereals all over Italy, in the valleys, plains, hills, and mountains. "You have all traveled through many lands; have you seen any country more fully cultivated than Italy?"[4] Varro also commented that the widespread conversion of cultivated fields to pasture increased the need for imported food.

Writing in the first century AD, Lucius Junius Moderatus Columella thought the best soil required minimal labor to produce the greatest yields. In his view, fertile topsoil well suited for grain should be at least two feet thick. Cereals grew best on valley bottom soils, but grapes and olives could flourish on thinner hillslope soils. Rich, easily worked soils made grains the major cash crop along Italy's river valleys. Focused like his predecessors on maximizing production, Columella chastised large landowners who left fields fallow for extended periods.

Columella described two simple tests of soil quality. The easy way was to take a small piece of earth, sprinkle it with a little water, and roll it around. Good soil would stick to your fingers when handled and did not crumble when thrown to the ground. A more labor-intensive test involved analyzing the dirt excavated from a hole. Soil that would not settle back down into the hole was rich in silt and clay good for growing grains; sandy soil that would not refill the hole was better suited for vineyards or pasture. Although little is known about Columella himself, I learned a version of his first test in graduate school at UC Berkeley.

Roman agriculturists recognized the importance of crop rotation—even the best soils could not grow the same crops forever. Farmers would periodically let a piece of ground lie fallow, grow a crop of legumes, or raise a cover crop well suited for the local dirt. Generally, they left fields fallow every other year between cereal crops. As for plant nutrition, Romans understood that crops absorbed nutrients from the soil and recognized the value of manure to achieve the greatest yields from the soil and prevent its exhaustion. In line with Cato's advice to keep "a large dunghill," Roman farmers collected and stored manure from oxen, horses, sheep, goats, pigs, and even pigeons for spreading on their fields. They applied marl—crushed limestone—as well as ashes to enrich their fields. Varro recommended applying cattle dung in piles but thought bird droppings should be scattered. Cato recommended using human excrement if manure was unavailable. Columella even cautioned that hillside fields required more manure because runoff across bare, plowed fields would wash the stuff downslope. He also advised plowing manure under to keep it from drying out in the sun.

Above all else, Roman agronomists stressed the importance of plowing. Repeated annual plowing provided a well-aerated bed free of weeds. Varro recommended three plowings; Columella advised four. Stiff soils were plowed many times to break up the ground before planting. By the peak of the empire, Roman farmers used light wooden plows for thin easily worked soils, and heavy iron plows for dense soils. Most still plowed in straight lines with equal-size furrows. Just as in Greece, all that plowing slowly pushed soil downhill and promoted erosion, as runoff from each storm took its toll—slow enough to ignore in one farmer's lifetime, but fast enough to add up over the centuries.

Roman farmers plowed under fields of lupines and beans to restore humus and maintain soil texture. Columella wrote that a rotation of heavily manured beans following a crop of cereal could keep land under continuous production. He specifically warned against the damage that slave labor did to the land. "It is better for every kind of land to be under free farmers than under slave overseers, but this is particularly true of grain land. To such land a tenant farmer can do no great harm . . . while slaves do it tremendous damage."[5] Columella thought that poor agricultural practices on large plantations threatened the foundation of Roman agriculture.

Caius Plinius Secundus, better known as Pliny the Elder (AD 23–79), attributed the decline of Roman agriculture to city-dwelling landlords leaving large tracts of farmland in the hands of overseers running slave labor. Pliny also decried the general practice of growing cash crops for the highest profit to the exclusion of good husbandry. He maintained that such practices would ruin the empire.

Some contemporary accounts support the view that the Romans' land use greatly accelerated erosion despite their extensive knowledge of practical husbandry. Pliny described how forest clearing on hillslopes produced devastating torrents when the rain no longer sank into the soil. Later, in the second century, Pausanias compared two Greek river basins: the Maeander, actively plowed agricultural land, and the Achelous, vacant land whose inhabitants had been removed by the Romans. The populated, actively cultivated watershed produced far more sediment, its rapidly advancing delta turning islands into peninsulas. But by how much did Roman agriculture increase erosion rates in Roman Italy?

In the 1960s Princeton University geologist Sheldon Judson studied ancient erosion in the area around Rome. He saw how the foundation of a cistern built to hold water for a Roman villa around AD 150 stood exposed by between twenty and fifty-one inches of erosion since the structure was built, an average rate of more than an inch per century. He noted a similar rate for the Via Prenestina, a major road leading west from Rome. Originally placed flush with the surface of the ridge along which it runs, by the 1960s its basalt paving stones protruded several feet above easily eroded volcanic soil of the surrounding cultivated slopes. Other sites around Rome recorded an average of three-quarters of an inch to four inches of erosion each century since the city's founding.

Sediments accumulated in volcanic crater lakes in the countryside confirmed the account of dramatic erosion. Cores pulled from Lago di Monterosi, a small lake twenty-five miles north of Rome, record that land shedding sediment into the lake eroded about an inch every thousand years before the Via Cassia was built through the area in the second century BC. After the road was built, erosion increased to almost an inch per century as farms and estates began working the land to produce marketable crops. Sediments from a lake in the Baccano basin, less than twenty miles north of Rome along the Via Cassia, also recorded an average erosion rate on the surrounding lands of a little more than an inch every thousand years for more than five thousand years before the Romans drained the lake in the second century BC. Thick deposits of material stripped from hillslopes and deposited in valley bottoms along streams north of Rome further indicate intense erosion near the end of the empire.

These diverse lines of evidence, together with Vita-Finzi's findings, point to a dramatic increase in soil erosion owing to Roman agriculture. Considered annually, the net increase was small, just a fraction of an inch per year—hardly enough to notice. If the original topsoil was six inches to a foot thick, it probably took at least a few centuries but no more than about a thousand years to strip topsoil off the Roman heartland. Once landowners no longer worked their own fields, it is doubtful that more than a handful noticed what was happening to their dirt.

It was easier to see evidence of soil erosion downstream along the major rivers, where ports became inland towns as sediments derived from soil stripped off hillsides pushed the land seaward. Swamped by sediment from the Tiber River, Rome's ancient seaport of Ostia today stands miles from the coast. Other towns, like Ravenna, lost their access to the sea and declined in influence. At the southern end of Italy, the town of Sybaris vanished beneath dirt deposited by the Crathis River.

Historians still debate the reasons behind the collapse of the Roman Empire, placing different emphases on imperial politics, external pressures, and environmental degradation. But Rome did not so much collapse as consume itself. While it would be simplistic to blame the fall of Rome on soil erosion alone, the stress of feeding a growing population from deteriorating lands helped unravel the empire. Moreover, the relation worked both ways. As soil erosion influenced Roman society, political and economic forces in turn shaped how Romans treated their soil.

When Hannibal razed the Italian countryside in the Second Punic War (218 to 201 BC), thousands of Roman farmers flooded into the cities as their fields and houses were destroyed. After Hannibal's defeat, vacant farmland was an attractive investment for those with money. The Roman government also paid off war loans from wealthy citizens with land abandoned during the war. The estimated quarter of a million slaves brought back to Italy provided a ready labor supply. After the war, all three of the primary sources of agricultural production—land, labor, and capital—were cheap and available.

The growth of large cash crop–oriented estates *(latifundia)* harnessed these resources to maximize production of wine and olive oil. By the middle of the second century BC, such large slave-worked plantations dominated Roman agriculture. The landowning citizen-farmer became an antiquated ideal but a convenient emblem for the Gracchi brothers' popular cause in 131 BC. They promoted laws giving a few acres of state-owned land to individual farmers, yet many of those who received land under the Gracchi laws could not make a living, sold their land off to larger landowners, and went back on the dole in Rome. Less than two centuries after the Gracchi brothers were assassinated, huge estates accounted for nearly all the arable land within two days' travel from Rome. Forbidden to engage directly in commerce, many wealthy senators circumvented the law by operating their estates as commercial farms. The total area under Roman cultivation continued to expand as the transformation from subsistence farming to agricultural plantations reshaped the Italian countryside.

The land fared poorly under these vast farming operations. In the first decade AD the historian Titus Livius wondered how the fields of central Italy could have supported the vast armies that centuries before had fought against Roman expansion—given the state of the land, accounts of Rome's ancient foes no longer seemed credible. Two centuries later Pertinax offered central Italy's abandoned farmland to anyone willing to work it for two years. Few took advantage of his offer. Another century later Diocletian bound free farmers and slaves to the land they cultivated. A generation after that, Constantine made it a crime for the son of a farmer to leave the farm where he was raised. By then central Italy's farmers could barely feed themselves, let alone the urban population. By AD 395 the abandoned fields of Campagna were estimated to cover enough land to have held more than 75,000 farms in the early republic.

The countryside around Rome had fed the growing metropolis until late in the third century BC. By the time of Christ, grain from the surrounding land could no longer feed the city. Two hundred thousand tons of grain a year were shipped from Egypt and North Africa to feed the million people in Rome. Emperor Tiberius complained to the Senate that "the very existence of the people of Rome is daily at the mercy of uncertain waves and storms."[6] Rome came to rely on food imported from the provinces to feed the capital's unruly mobs. Grain was shipped to Ostia, the closest port to Rome. Anyone delaying or disrupting deliveries could be summarily executed.

North African provinces faced constant pressure to produce as much grain as possible because political considerations compelled the empire to provide free grain to Rome's population. The Libyan coast produced copious harvests until soil erosion so degraded the land that the desert began encroaching from the south. The Roman destruction of Carthage in 146 BC, and its salting of the surrounding earth to prevent its resurrection are well known. Less widely appreciated are the longer-term effects of soil degradation when the growing Roman demand for grain reintensified cultivation in North Africa.

The Roman Senate paid to translate the twenty-eight volumes of Mago's handbook of Carthaginian agriculture salvaged from the ruined city. Once the salt leached away, land-hungry Romans turned the North African coast into densely planted olive farms—for a while. Major farming operations centered around great olive presses developed in the first century AD. Proconsuls charged with producing food for Rome commanded legions of up to two hundred thousand men to protect the harvest from marauding nomads. The barbarians were kept at bay for centuries, but the threat of soil erosion was harder to stop as political stability under the *pax romana* encouraged continuous cultivation aimed at maximizing each year's harvest. By the time the Vandals crossed from Spain into Africa and took Carthage in AD 439, the Roman presence was so feeble that fewer than fifteen thousand men conquered all of North Africa. After the Roman capitulation, overgrazing by herds of nomadic sheep prevented rebuilding the soil.

Today we hardly think of North Africa as the granary of the ancient world. Yet North African grain had relieved the Greek famine in 330 BC, and Rome conquered Carthage in part to secure its fields. The Roman Senate annexed Cyrenaica, the North African coast between Carthage and Egypt, in 75 BC, a year when war in Spain and a failed harvest in Gaul meant that the northern provinces could barely feed themselves, let alone the capital. With hungry rioters in Rome, it is likely that the Senate annexed Cyrenaica for its ability to produce grain.

Evidence for extensive ancient soil erosion in the region challenges the idea that a shifting climate caused the post-Roman abandonment of irrigated agriculture in North Africa. Although much of North Africa controlled by Rome was marginal agricultural land, archaeological evidence reported from UNESCO surveys in the mid-1980s confirms the record of initial Roman colonization by farmers in individual, self-sufficient households. Over the next few centuries, irrigated agriculture had gradually expanded as farms coalesced into larger operations focused on growing grains and olives for export.

The first Christian to write in Latin, Quintus Septimius Florens Tertullianus (known today simply as Tertullian), lived in Carthage around AD 200. In describing the end of the Roman frontier in North Africa he warned of overtaxing the environment. "All places are now accessible, well known, open to commerce. Delightful farms have now blotted out every trace of the dreadful wastes; cultivated fields have overcome woods. . . . We overcrowd the world. The elements can hardly support us. Our wants increase and our demands are keener, while Nature cannot bear us."[7]

The UNESCO archaeological surveys described evidence that helps explain Tertullian's unease as increased population densities led to widespread erosion on slopes beyond the small floodplains along rivers and streams. Pressure to defend a frontier with limited water and vanishing soil had gradually transformed Roman agricultural settlements in Libya into massive fortified farms staked out every few hundred feet along valley bottoms. The region was no longer a prosperous agricultural center by the time that ʿAmr ibn al ʿAṣ overran the remnants of Byzantine colonial authority in the seventh century.

In 1916 Columbia University professor Vladimir Simkhovitch argued that lack of dirt caused the decline of the Roman Empire. Soil exhaustion and erosion had depopulated the Roman countryside in the empire's late days; he pointed out that the amount of land needed to support a Roman farmer had increased from the small allotment given to each citizen at the founding of Rome to ten times as much land by the time of Julius Caesar. Simkhovitch noted how the philosopher Lucretius in his epic poem *De rerum natura* echoed other contemporary sources referring to the declining productivity of Mother Earth.

Did the perception of a general decline in soil fertility in the Roman heartland accurately reflect reality? That's hard to say. Columella, writing around AD 60, addresses the issue in a preface to *De re rustica*. "I frequently hear the most illustrious men of our country complaining that the sterility of our soil and intemperate weather have now for many ages past been diminishing the productivity of the land. Others give a rational background to their complaints, claiming that the land became tired and exhausted from its productivity in the former ages."[8]

Columella goes on to state that previous agricultural writers, most of whose works have not survived, uniformly complained of soil exhaustion. But soil need not inevitably succumb to old age, worn out by long cultivation. Instead, Columella argues, since the gods endowed soil with the potential for perpetual fecundity, it was impious to believe it could become exhausted. He qualifies his words, though, offering the opinion that soil would retain its fertility indefinitely if properly cared for and frequently manured.

Why begin a practical guide to farmers this way? In casting his argument as he did, Columella was pointing out that Rome's agricultural problem did not reflect some natural process of universal decay but rather Roman farmers' treatment of their land. Their problems were of their own making. In the second century BC Varro had referred to abandoned fields in Latium, describing an example of notoriously sterile soil where meager foliage and starved vines struggled to survive on land that had supported families a few centuries earlier. Taking up the point several centuries later, Columella insisted that the people of Latium would have starved without a share of the food imported to feed the capital.

Some historians see the growing indebtedness of Roman farmers as contributing to the empire's internal turmoil. Farm debt can come from borrowing to provide the tools necessary to run the farm or because farm income fails to meet the needs of the farmer's family. The low capital requirements of Roman farming suggest that farmers working the republic's traditional small farms were having a hard time feeding themselves. Large estate owners took advantage of their distressed neighbors and bought up huge tracts of land. Contrary to the conventional wisdom that civil strife and wars had depopulated the Roman countryside, the disappearance of small farms occurred during a period of unprecedented peace. Roman laws prohibiting the separation of agricultural slaves from the land they worked passed in response to the abandonment of the Roman countryside. Eventually the problem became so acute that even free tenant farmers were decreed tied to the soil they plowed—and thus to the land's owners. The social arrangement between farmer-serfs and landowning nobles established by these laws survived long after the empire crumbled (to set the stage, many historians believe, for medieval serfdom).

Still, how could Italy's soil decline when Romans knew about agricultural husbandry, crop rotations, and manure? Such practices require using a portion of the farmer's income to improve the soil, whereas maximizing its immediate yield involves cashing in soil fertility. In addition, farmers sinking into debt or hunger understandably push to get everything they can from even degraded soil.

To some degree, the Roman imperative to acquire land was driven by the need to secure food for the growing populace. Viewed in this context, the shrinking harvests from central Italy's fields encouraged intensive agriculture in newly conquered provinces. Soil erosion progressively degraded the Roman heartland and then spread to the provinces—except Egypt, which became a colony exploited to feed Rome upon the death of Cleopatra in 30 BC.

Egypt remained immune because of the life-giving floods of the Nile. The Nile's importance to the Roman Empire is clear in Egypt's status as the emperor's personal possession. In the first century AD Emperor Augustus forbade senators or Roman nobility to enter Egypt without his permission. "Whoever made himself master of Alexandria . . . might with a small force . . . reduce all Italy to a famine."[9] By the end of the empire, the dirt of the Nile fed Rome. Soil erosion alone did not destroy Rome, but the dirt of modern Italy and former Roman colonies speaks for itself.

More than a thousand years after the fall of Rome, a widely traveled New England lawyer explored the role of soil erosion on ancient societies. Born in 1801 in the frontier community of Woodstock, Vermont, George Perkins Marsh traveled extensively through the Old World and published *Man and Nature,* the foundational work of environmentalism in 1864. Marsh was a voracious reader who gave up law to run for Congress in 1843 and was appointed U.S. minister to Turkey five years later. With minimal duties and ample time for travel, he collected plants and animals for the Smithsonian Institution during an expedition through Egypt and Palestine in 1851 before returning home. A decade later President Abraham Lincoln appointed Marsh ambassador to Italy. Extending his travels to the Alps, Marsh saw the Old World's degraded land as the end result of soil neglect like that he witnessed as Vermont's forests were converted into wheat fields and pastures.

> Territory larger than all Europe, the abundance of which sustained in bygone centuries a population scarcely inferior to that of the whole Christian world at the present day, has been entirely withdrawn from human use, or, at best, is thinly inhabited. . . . There are parts of Asia Minor, of Northern Africa, of Greece, and even of Alpine Europe, where the operation of causes set in action by man has brought the face of the earth to a desolation almost as complete as that of the moon; and though, within that brief space of time which we call "the historical period," they are known to have been covered with luxuriant woods, verdant pastures, and fertile meadows.[10]

Marsh's revelation was twofold: land did not necessarily recover after being abused, and people destroyed the balance of nature unconsciously in the pursuit of more immediate ends. Calling attention to how unintended consequences of human activity affected the ability of the land to support societies, he was confident that America could prevent repeating the Old World's folly.

Marsh believed agricultural technology could keep up with the world's rising population—as long as there was soil left to plow. But he recognized that the capacity for damaging land also increased with technological sophistication. Popularizing the notion that deforestation and soil erosion ruined civilizations in the Middle East, Marsh challenged American confidence in the inexhaustibility of resources. His book became an instant classic, supporting three editions that he worked on revising literally until the last day of his life.

Half a century later, on the eve of the Second World War, the U.S. Department of Agriculture sent renowned soil expert Walter Lowdermilk to survey the effects of land use on erosion in the Middle East, North Africa, and Europe. Hitler's invasion of Poland prevented him from continuing on to central Europe and the Balkans, but like Marsh before him Lowdermilk had already seen enough of Europe and Asia to consider the Old World's soils a graveyard of empires.

Visiting the ancient Roman agricultural colonies in Tunisia and Algeria on the northern coast of Africa, Lowdermilk found a landscape Cato would hardly have considered a threat. "Over a large part of the ancient granary of Rome we found the soil washed off to bedrock and the hills seriously gullied as a result of overgrazing. Most valley floors are still cultivated but are eroding in great gullies fed by accelerated storm runoff from barren slopes."[11] At Cuicul, the ruins of a great city lay under three feet of dirt washed off rocky slopes formerly covered with olive plantations. A few remnant groves remained perched on soil pedestals a foot or two above bedrock slopes.

The ancient city of Timgad impressed Lowdermilk even more. Founded by Trajan at the height of Roman power in the first century AD, the city supported a large public library, a 2,500-person theater, more than a dozen bathhouses, and public bathrooms with marble flush toilets. Lowdermilk found the city housed a few hundred inhabitants living in stone buildings recycled from the ancient ruins. Abandoned for more than a thousand years, the remains of giant olive presses standing on bare, treeless slopes spoke of better days.

In Tunisia, Lowdermilk pondered the ruins of a 60,000-seat amphitheater, in its day second in size only to Rome's coliseum. He estimated the modern population of the surrounding district to be less than a tenth of the amphitheater's capacity. Could a drying climate have forced people to abandon their fields to the desert? Lowdermilk doubted the conventional proposition. The director of archaeological excavations at Timgad had grown an olive grove using Roman methods on an unexcavated portion of the valley. Its good health showed that the climate had not changed enough to explain the region's agricultural decline.

At Sousse, Lowdermilk found live olive trees thought to be 1,500 years old, confirming that a drying climate was not responsible for the collapse of North African agriculture. Soil remained on the slopes in these ancient

Figure 9. Ruins of the first century A.D. Roman city of Timgad, North Africa (Lowdermilk 1953, 17, fig. 9).

groves, held in place by ancient terraces and banks of earth that guided runoff from the surrounding slopes to the fields. Hills protected from grazing retained soil covered by grass and scattered trees. Concluding that elsewhere the soil had been washed from the slopes, Lowdermilk blamed overgrazing for unleashing erosion that destroyed the capacity of the land to support people.

Traveling east, Lowdermilk's party reached the area where Moses led the Israelites out of the desert and into the Jordan Valley. Stopping at Jericho, Lowdermilk found the red soil had been stripped from more than half the upland area. Deep gullies sliced through valley bottoms, the remnants of which were still being farmed. More than three-quarters of the ancient villages on steep slopes had been abandoned, while nine out of ten on the valley bottom remained inhabited. Abandoned villages stood where the soil had been stripped away. Thick soil remained on cultivated slopes where stone terraces were well maintained.

At Petra, the capital of the Nabataean civilization carved from bedrock on the edge of the Arabian Desert around 200 BC, Lowdermilk saw more bare rocky slopes covered with ruined terraces. Pondering where the soil once held by the terraces had gone, he concluded that the city ate food grown on nearby slopes until invading nomads triggered a breakdown in soil conservation measures. Today the great theater that could seat thousands entertains few tourists.

Crossing into Syria, Lowdermilk visited the ruined Roman city of Jerash. Home to a quarter of a million people in biblical times, the ancient city lies beneath more than ten feet of dirt washed off the surrounding slopes. Contrary to archaeologists' then-favored theory, Lowdermilk found no evidence that the water supply had failed. Just like those at Petra and Jericho, rock-walled terraces once retained soil on slopes now stripped down to bedrock. What was left of the original soil lay trapped in the valley bottoms. Home to just a few thousand people in the 1930s, the area once supported luxurious villas that sent boatloads of grain to Rome.

Continuing north, Lowdermilk reached Antioch where Paul began preaching the Gospel to the region's largest supply of potential converts. A hundred villages and towns surrounded the greatest and richest city of ancient Syria during the Roman occupation. Only seven villages remained inhabited in the 1970s. Four times as many lay beneath a marshy swamp built up by silt eroded off hillsides under aggressive Roman cultivation. Archaeologists had to dig down twenty-eight feet to uncover parts of the ruined city.

They also found clues as to where all the dirt came from. Most telling were rough, unfinished parts of foundations not intended to be seen and doorsills lacking steps stranded three to six feet above bare rocky ground in the uplands north of Antioch. With the soil long gone, the region once famous for exporting grain and olive oil now supports a few seminomads. Lowdermilk described the region's ruins as stark stone skeletons rising high above the bare rock slopes of a desert its inhabitants inadvertently made, leaving an enduring legacy of impoverishment. Whatever else may have happened in the region, Syria's upland soil was gone.

Lebanon's dirt met a similar fate. About 4,500 years ago the Phoenicians moved west from the desert to the eastern shore of the Mediterranean. Starting on a narrow strip of land on the coastal plain, they brought Mesopotamian agricultural practices to their new home, a land of cedar forests with little flat land to cultivate. After plowing up the narrow coastal plain, Phoenicians cleared sloping fields and sold the timber to their treeless neighbors in Mesopotamia and Egypt. Whether the great cedar forest was cut primarily for timber or farmland, the two went hand in hand as farms spread up the slopes.

Pounded by heavy winter rains, soil eroded rapidly from plowed hillsides. Adapting their agricultural methods to this new problem, Phoenicians began terracing to retain soil. Lowdermilk described how the ruined walls of ancient terraces lay scattered throughout the region. Few agricultural practices are as labor-intensive as terracing—especially on steep slopes. Effective at slowing erosion if maintained, terraces rapidly lose effectiveness if neglected.

But only a fraction of Lebanon's steepest land was terraced. Storm by storm, soil eroded off most slopes. By the ninth century BC Phoenician emigrants began colonizing North Africa and the western Mediterranean, sending manufactured goods to distant outposts in exchange for food. By the time Alexander the Great conquered Lebanon in 322 BC, Phoenicia's golden age was over. Cut off from the colonies, and with most of the homeland's topsoil gone, Phoenician civilization never recovered.

By the time of Lowdermilk's visit four small cedar groves were all that remained of the great cedar forest that covered two thousand square miles of ancient Phoenicia. Seeing that these remnant groves grew in patches of valley bottom soils protected from goats convinced Lowdermilk that the forest did not disappear because the climate changed. The big trees could not grow back once the soil was gone, and overgrazing kept it from rebuilding.

In the summer of 1979 radiocarbon dating of sediment cores recovered from the Sea of Galilee (Lake Kinneret) showed that the erosion rate from the surrounding land more than doubled around 1000 BC. This time corresponds to increased settlement and agricultural expansion into mountainous areas following the Israelites' arrival. Pollen preserved in the different layers of lake sediments show that between the founding of the kingdom of Israel and the end of the Roman occupation about thirteen hundred years later, olives and grapes replaced the native oak forest.

When Moses led the Israelites out of the desert into Canaan, they appeared to have arrived in an agricultural paradise. "For the Lord thy God bringeth thee into a good land, a land of brooks of water, of fountains and depths that spring out of valleys and hills; a land of wheat and barley and vines and fig trees and pomegranates, a land of olive oil and honey" (Deuteronomy 8:7–8). Inconveniently, however, the best valley-bottom land was already occupied.

When Moses and company arrived, Canaan was a collection of city-states subjugated by Egyptian military superiority. Heavily fortified Canaanite cities controlled the agricultural lowlands. Undeterred, the new arrivals farmed the vacant uplands. "But the mountain-country shall be yours. Although it is wooded, you shall cut it down, and its farthest extent shall be yours" (Joshua 17:18). Settling in small villages, they cleared the forests and farmed terraced slopes in the hill country to gain a foothold in the Promised Land.

The Israelites adopted traditional Canaanite agriculture in their new hillside farms, growing what their neighbors grew. But they also practiced crop rotation and fallowing and designed systems to collect and deliver rainwater to their terraced fields. With the development of new iron tools, greater harvests led to agricultural surpluses that could support larger settlements. Leaving fields fallow every seventh year was mandated and animal dung was mixed with straw to produce compost.[12] Land was regarded as God's property entrusted to the people of Israel for safekeeping. In the Judean highlands, Lowdermilk noted how a few well-maintained stone terraces still held soil after several thousand years of cultivation.

Agriculture expanded so much under the later Roman occupation that the empire's Middle East provinces were completely deforested by the first century AD. Grazing typically replaced forests on terrain too steep to farm. Throughout the region, flocks of goats and sheep reduced vegetation to stubble. Catastrophic soil erosion followed when too many livestock grazed steep hillsides. Forest soils built up over millennia disappeared. Once the soil was gone, so was the forest.

In a radio address from Jerusalem in June 1939, Walter Lowdermilk offered an eleventh commandment he imagined Moses might have slipped in had he foreseen what was to become of this promised land. "Thou shalt inherit the Holy Earth as a faithful steward, conserving its resources and productivity from generation to generation. Thou shalt safeguard thy fields from soil erosion . . . and protect thy hills from overgrazing by thy herds, that thy descendants may have abundance forever. If any shall fail in this stewardship of the land . . . thy descendants shall decrease and live in poverty or perish from off the face of the Earth."[13]

As Marsh had before him, Lowdermilk worried about the implications of what he saw in the Middle East for America's long-term prospects. Both looked to the Old World for lessons for the New World. Neither realized that the scenarios they worried about had already happened in America.

Mayan civilization provides the best-studied, but by no means the only, example of soil degradation contributing to the collapse of a society in the Americas. The earliest Mayan settlements grew from the lowland jungle of the Yucatán Peninsula and slowly consolidated into increasingly complex settlements. By the second century BC, large ceremonial and commercial centers like Tikal coalesced into a complex hierarchical society of city-states with a common language, culture, and architecture. Mayan cities were comparable in size to Sumerian city-states. At its peak Tikal was home to from thirty thousand to fifty thousand people.

The first settled Mesoamerican communities grew to regional prominence after maize was domesticated about 2000 BC. Over the next thousand years, small villages came to depend on cultivating maize to supplement hunting and gathering from the wild lands between villages. Small-scale forest clearing for agriculture gradually expanded and maize became an increasingly important component of the Mesoamerican diet. As in Mesopotamia, dispersed networks of settlements grew into ceremonial centers and towns with priests, artisans, and administrators to oversee the redistribution of surplus food.

Initially, the productivity of domesticated maize was similar to wild varieties, which could be readily gathered. Tiny cobs about the size of a human thumb were simply chewed. People began grinding maize into flour once high-yield varieties allowed the development of permanent settlements. The region supported a diffuse rural population until large towns emerged between 350 BC and AD 250. By then some parts of the Mayan world were already severely eroded, but in many areas the greatest soil erosion—and evidence for soil conservation efforts—date from about AD 600–900. The subsequent lack of artifacts has been interpreted to show a dramatic population decline (or dispersal) as Mayan society crumbled and the jungle reclaimed Tikal and its rivals.

Mayan population grew from less than two hundred thousand in 600 BC to more than a million by AD 300. Five hundred years later at the peak of Mayan civilization, the population reached at least three million and perhaps as many as six million. Over the next two hundred years the population fell to less than half a million people. When John Stephens rediscovered ruined Mayan cities the region appeared deserted except for areas on the edge of the jungle. Even today the population density in the rapidly growing region remains below that of ancient times. So what happened?

Mayan agriculture began with a system known as slash-and-burn agriculture, in which a patch of jungle was cleared with stone axes and then burned before the onset of the rains, when maize and beans were planted. Ash from burning the cleared forest fertilized the soil and guaranteed good crops for a few years, after which fertility of the nutrient-poor tropical soil fell rapidly. Cleared patches could not be farmed for long before being abandoned to the jungle to restore soil fertility. A lot of jungle was needed to keep a few fields under cultivation. As in ancient Greece and Italy, the first evidence for extensive soil erosion coincides with pioneer farming.

Slash and burn worked well while the population density remained low and there was enough land for farmers to move their fields every few years. As the great Mayan cities rose from the jungle, people kept clearing land as their ancestors had done, but they stopped moving their fields. The tropical soils of the Yucatán Peninsula are thin and easily eroded. Under sustained cultivation, the high productivity obtained right after clearing and burning rapidly declines. Compounding this problem, the lack of domesticated animals meant no manure for replenishing the soil. Just as in Greece and Rome, rising demand for food and declining productivity compelled cultivation of increasingly marginal land.

After about 300 BC the region's increasing population led people to begin farming poorly drained valley bottoms and limestone slopes with thin, fragile soils. They built raised fields in swamps by digging networks of drainage canals and piling up the excavated material in between to create raised planting beds perched above the water table. In some areas extensive terracing began around AD 250 and then spread across the landscape as the population continued to expand for another six and a half centuries. Mayan farmers terraced hillsides to create flat planting surfaces, slow erosion by overland flow, and divert water to fields. However, in major areas like Tikal and Copán there is little evidence for soil conservation efforts. Even with erosion control efforts, deposition of soil eroded from surrounding slopes disrupted wetland agriculture practiced in sinkholes.

Sediment cores from lakes in the Mayan heartland suggest that agricultural intensification increased soil erosion. The rate that sedimentation piled up on lakebeds increased substantially from 250 BC through the ninth century AD. While not necessarily responsible for the collapse of Mayan society, soil erosion peaked shortly before Mayan civilization unraveled about AD 900 when the food surpluses that sustained the social hierarchy disappeared. Some Mayan cities were abandoned with buildings half finished.

In the 1990s geographers studying small depressions, known as *bajos,* around Mayan sites in northwestern Belize found that cultivated wetlands had filled with soils eroded after deforestation of the surrounding slopes. The southern Yucatán is broken into depressions that formed natural wetlands extensively cultivated during the peak of Mayan civilization. Trenches revealed buried soils dating from the pre-Mayan period covered by two and a half to six feet of dirt eroded from the surrounding slopes in two distinct episodes. The first corresponded to forest clearing during the spread of pioneer farmers from the valleys up onto the surrounding hillslopes. The second occurred during agricultural intensification immediately before the end of Mayan civilization, after which soil began to rebuild as the forest reclaimed fields and wetlands.

Researchers also found evidence for accelerated soil erosion caused by extensive deforestation of sloping land in the Mayan lowlands. Where Mayan terraces remain intact, they hold three to four times more soil than lies on adjacent cultivated slopes. Development of erosion control methods allowed the Mayan heartland to support large populations but the expansion depended on intensive cultivation of erosion-prone slopes and sedimentation-prone wetlands. Eventually, Mayan civilization reached a point where its agricultural methods could no longer sustain its population.

Modern deforestation in the Petén is beginning to repeat the cycle of erosion after a thousand years of soil development. Since the early 1980s landless peasant farmers have turned much of the region's forest into traditional Mayan *milpas* (small cultivated fields). A twentyfold increase in population from 1964 to 1997 has transformed the region from nearly unbroken forest to a nearly deforested landscape.

Soils on most of the region's hillslopes consisted of an organic horizon above a thin mineral soil sitting directly on weakly weathered limestone bedrock. One study found that under the region's last virgin forest, hillslope soils were about ten to twenty inches thick, whereas modern cultivated fields are already missing three to seven inches of topsoil—most of the O and A horizons. In some places, the rapid erosion following modern slope clearing and cultivation had already stripped the soil down to bedrock. Another study of soil erosion following modern forest clearance in central Belize found that one to four inches of topsoil were lost in a single ten-year cycle of clearing for corn and cassava for two to three years of cultivation followed by a fallow year. Complete removal of the soil would require just four milpa cycles. In a particularly striking example from the northern Yucatán, about eight inches of soil on the upland surfaces around a sinkhole named Aguada Catolina was eroded to bedrock in a decade of renewed cultivation. Similarly, researchers investigating ancient soil erosion in the Mayan of the Petén noticed that the soil was stripped down to bedrock on newly cleared slopes in under a decade.

Rates of soil formation in the Central American jungle are far slower than rates of erosion under Mayan agriculture. The region's limestone bedrock weathers about half an inch to five inches in a thousand years. An average soil depth of about three inches developed on Mayan architecture abandoned a thousand years ago indicates rates of soil formation similar to the geologic erosion rate. Both are about a hundred times slower than erosion from cultivated slopes.

The Mayan heartland was not the only place where soil influenced Native American civilizations. Soils of central Mexico tell similar stories of severe erosion on steep hillslopes undermining agriculture.

In the late 1940s UC Berkeley professor Sherburne Cook drove around the central Mexican plateau and concluded that the land was in poorest condition in areas that had supported the largest populations before the Spanish conquest. The thick soil and sod covering uncultivated areas contrasted with the truncated soil profiles, slopes stripped down to weathered rock, and thick, artifact-rich valley fills derived from former hillslope soils that characterized more densely populated areas. Cook saw evidence for two periods of erosion, an ancient episode that stripped soils from hillsides and a more recent episode that entrenched deep gullies into the valley bottoms. "Evidently the entire range was once cleared for cultivation, abandoned, allowed to become covered with young forest, and finally, the lower portion again cleared."[14] Despite Cook's revelation, the timing of these cycles remained uncertain until development of radiocarbon dating in the 1950s.

Analyses of sediment cores from Lake Pátzcuaro, east of Mexico City in Michoacán, revealed evidence for three distinct periods of rapid soil erosion. The first period accompanied extensive land clearance about 3,500 years ago shortly after maize cultivation began. The second period of high erosion occurred in the late preclassic period between 2,500 and 1,200 years ago. The third erosional period peaked immediately before the Spanish conquest, when up to a hundred thousand people lived around the lake. Despite the introduction of the plow, soil erosion rates dropped as diseases decimated the region's population after Cortez arrived in AD 1521.

Just as in ancient Greece and around the Mediterranean, cycles of soil erosion in different parts of central Mexico occurred at different times and so were not driven by changes in climate. For example, in the Puebla-Tlaxcala area of the central Mexican highland, accelerated erosion of hillslope soils around 700 BC coincided with rapid expansion of settlements. A period of soil formation and cultural stagnation beginning about AD 100 was followed by a second period of accelerated erosion and rapid expansion of settlements before the Spanish conquest. Geoarchaeological surveys of hillslopes in the region revealed that agricultural fields were abandoned owing to soil erosion progressively from top to bottom, much as in ancient Greece.

As in the *bajos* of the Mayan jungle, pits dug into the swampy valley bottom sediments in the Upper Lerma Basin in central Mexico also record increased soil erosion from the surrounding slopes beginning around 1100 BC. Soil erosion then intensified during expansion of settlements in the late classic and early postclassic periods beginning about AD 600. On his tour fourteen centuries later, Cook recognized that the areas populated most densely in preconquest times had the worst soil exhaustion.

The soils in Mexico's "cradle of maize" in the Tehuacán Valley about a hundred miles southeast of Mexico City also bear witness to extensive preColumbian soil erosion. In the early 1990s field surveys of soils around the town of Metzontla revealed striking differences between areas that had been cultivated and those that had not. Intensively cultivated hillslopes were extensively eroded, with thin soils above weathered rock. Remnants of subsoil exposed at the ground surface documented soil erosion from fields, leaving a modern soil that consists of little more than a thin mantle of broken rock. In contrast, areas with little evidence of past cultivation contained a foot and a half of well-developed soil above weathered rock. Abrupt transitions in soil depth between long-cultivated and uncultivated areas suggest that a foot and a half of soil was missing from the farms.

Expansion of agriculture from irrigated valley bottoms up onto the surrounding hillslopes about 1,300 to 1,700 years ago supported a growing population and triggered widespread soil erosion that still impoverishes the region. Agriculture on the slopes of the region today supplies only about a quarter of the small town's maize and beans. Metzontla's residents produce handcrafted ceramics or work at wage labor in other towns. In this semiarid area where soil production proceeds slowly, the residents' primary environmental concern is access to firewood for domestic use and for firing ceramics. Their soil disappeared slowly enough that they don't know it is gone.

Erosion from agriculture also caused abandonment of parts of southern Central America. Pollen from a long core pulled from the bottom of La Yeguada's small lake in central Panama records that the rainforest was cleared for slash-and-burn agriculture between 7,000 and 4,000 years ago.

Archaeological records from this period indicate considerable population growth as intensified agriculture stripped the forest from the lake's watershed. By the time of Christ, accelerated erosion in the foothills and uplands led to agricultural abandonment of the watershed. Slow forest regeneration suggests depleted soils, and later agricultural settlements were concentrated along previously unoccupied floodplains and coastal valleys. The uppermost layers in the long sediment core revealed that the primeval rainforest actually dates from the time of the Spanish conquest, when the indigenous population of the area again declined dramatically—this time felled by disease.

In the American Southwest, the spectacular ruins of Mesa Verde, Chaco Canyon, and Canyon de Chelly—all abandoned well before discovery by Euro-Americans—have long intrigued archaeologists. Between about AD 1250 and 1400, the native Pueblo culture vanished from the Southwest. The usual suspects of war, disease, drought, and deforestation have been proposed to explain the mystery.

Pollen sequences recovered from different depths in valley bottom sediments show little to no change in the vegetation community at Chaco Canyon for thousands of years—until the Pueblo people arrived. Plant remains preserved in crystallized packrat urine built up on the floor of caves show that the native vegetation was pinyon-juniper woodland, and that the local vegetation changed dramatically during Pueblo occupation. The inhabitants of Chaco Canyon used thousands of ponderosa pines to construct buildings between AD 1000 and 1200. Countless more trees were burned as fuel. Today the local vegetation on most of the valley floor is a mix of desert scrub and grasses. But if you hike near the canyon you can still see ancient stumps in areas where few trees now grow.

Many have argued that droughts led to the abandonment of Chaco Canyon. Although droughts probably contributed to the Pueblo culture's decline, the regional climate for the past thousand years falls within the range of variability for the past six thousand years. It seems more likely that salinization of the Pueblos' fields and soil erosion limited the life span of their agriculture as a growing population led to dependence on neighboring areas for basic resources. These conditions set up an agricultural disaster during the next drought.

Domesticated maize arrived at Chaco Canyon about 1500 BC. Initially grown near ephemeral streams or freshwater marshes, maize production increasingly depended on floodplain irrigation as agriculture expanded. By about AD 800 to 1000, rain-fed farming was practiced wherever feasible throughout the Southwest. Agricultural settlements ranged in size from small communities of a few dozen people to villages with hundreds of inhabitants. Foraging remained an important component of the diet—particularly during droughts.

At first sites were occupied for a few decades before people moved on to new locations, but by about AD 1150 there was no unused arable land to move into or cultivate when local crops failed. The landscape was full, the desert's rainfall was capricious, and its soil was fragile. As in the Old World centuries earlier, settlements became increasingly sedentary and their heavy investment in agricultural infrastructure discouraged farmers from leaving fields fallow every few years. Beginning about AD 1130, two centuries of drought and chaotic rainfall patterns occurred while all the arable land was already under cultivation. When crop failures on marginal land forced people to move back into more settled areas, the remaining productive land could not support them.

Comparison of ancient agricultural soils and uncultivated soils in New Mexico and Peru shows that agricultural practices need not undermine societies. Soils at a site in the Gila National Forest, typical of prehistoric agricultural sites in the American Southwest, were cultivated between AD 1100 and 1150, at the peak of Pueblo culture, and subsequently abandoned. Soils of sites cultivated by the Pueblo culture are lighter colored, with a third to a half of the carbon, nitrogen, and phosphorus content of neighboring uncultivated soils. In addition, cultivated plots had gullies—some more than three feet deep—that began during cultivation. Even today, little grass grows on the ancient farm plots. Native vegetation cannot recolonize the degraded soil even eight centuries after cultivation ceased.

In contrast, modern farmers in Peru's Colca Valley still use ancient terraces cultivated for more than fifteen centuries. Like their ancestors, they maintain soil fertility through intercropping, crop rotations that include legumes, fallowing, and the use of both manure and ash to maintain soil fertility. They have an extensive homegrown system of soil classification and do not till the soil before planting; instead they insert seeds into the ground using a chisel-like device that minimally disturbs the soil. These long-cultivated soils have A horizons that are typically one to four feet thicker than those of neighboring uncultivated soils. The cultivated Peruvian soils are full of earthworms and have higher concentrations of carbon, nitrogen, and phosphorus than native soils. In contrast to the New Mexican example, under traditional soil management these Peruvian soils have fed people for more than fifteen hundred years.

The contrast between how the Pueblos and the Incas treated their dirt is but another chapter in the broader story of how the rise of agriculture set off a perpetual race to figure out how to feed growing populations by continuously increasing crop yields. Sometimes cultures figured out a way to muddle through without depleting soil productivity, often they did not.

A common lesson of the ancient empires of the Old and New Worlds is that even innovative adaptations cannot make up for a lack of fertile soil to sustain increased productivity. As long as people take care of their land, the land can sustain them. Conversely, neglect of the basic health of the soil accelerated the downfall of civilization after civilization even as the harsh consequences of erosion and soil exhaustion helped push Western society from Mesopotamia to Greece, Rome, and beyond.

Efforts to feed the world today often include calls for a cultural revolution, a new agrotechnological revolution, or a political revolution to redistribute land to subsistence farmers. Less widely known is how, after centuries of agricultural decline, a preindustrial agricultural revolution began in the still-fertile and revitalized fields of Western Europe, setting the stage for the social, cultural, and political forces that forged colonial powers and shaped our modern global society.

FIVE

Let Them Eat Colonies

There is nothing new except what has been forgotten.

MARIE ANTOINETTE

GUATEMALANS GROW SOME OF THE BEST COFFEE in the world, but most can't buy it at home. Neither can tourists. When I was there last I had to wake up on freeze-dried Mexican Nescafé, even though I can buy bags of freshly roasted Guatemalan coffee beans two blocks from my house in Seattle. Less well known than the story of how Europe carved out global empires is how the way Europeans treated their soil helped launch the exploration and history of the New World. Today's globalized agriculture that ships local produce overseas to wealthier markets reflects the legacy of colonial plantations established to help feed European cities.

Like many ancient agricultural societies, Europeans began working to improve their dirt once soil fertility and access to fresh land declined. But unlike the Mediterranean's intense spring and summer rains that promoted erosion from bare fields, western Europe's gentle summer rains and winter-spring snow pack limited erosion of even highly erodible loess soils when farmed. Moreover, by rediscovering soil husbandry western Europeans kept soil degradation and erosion at bay long enough to establish colonial empires that provided new land to exploit.

Farming spread from the Middle East into Greece and the Balkans between seven and eight thousand years ago. After moving into central Europe's easily worked loess, agriculture steadily advanced north and west, reaching Scandinavia about three thousand years ago. Consuming Europe's forest soils as it went, agriculture left a record of boom-and-bust cycles associated first with Neolithic and Bronze Age cultures, then Iron Age and Roman society, and most recently the medieval and modern periods when colonial empires began mining soil and sending both produce and profits back to feed Europe's increasingly urban populace—the Industrial Revolution's new class of landless peasants.

The first agricultural communities reached Europe's doorstep in southern Bulgaria around 5300 BC. At first farmers grew wheat and barley in small fields surrounding a few timber-framed buildings. Agricultural expansion into marginal land lasted about two thousand years before the agricultural potential of the region was fully exploited and persistent cultivation began to exhaust the soil. With no evidence of a climate shift, local populations grew and then declined as agricultural settlement swept through the area. Evidence for extensive late Neolithic soil erosion shows that agriculture spread from small areas of arable soils on the valley bottoms into highly erodible forest soils on steeper slopes. Eventually, the landscape filled in with small communities of several hundred people farming the area within about a mile of their village.

In these first European communities, population rose slowly before a rapid decline that emptied settlements out for five hundred to a thousand years, until the first traces of Bronze Age cultures then appeared. This pattern suggests a fundamental model of agricultural development in which prosperity increases the capacity of the land to support people, allowing the population to expand to use the available land. Then, having eroded soils from marginal land, the population contracts rapidly before soil rebuilds in a period of low population density.

This roller-coaster cycle characterizes the relation between population and food production in many cultures and contexts because the agricultural potential of the land is not a constant—both technology and the state of the soil influence food production. Improved agricultural practices can support more people with fewer farmers, but soil health eventually determines how many people the land can support. Floodplains continually receive nutrients from periodic flooding, but most other land cannot produce continuously high crop yields without intensive fertilization. So once a society comes to depend on upland farming it can cultivate a fraction of its land base at any one time, expand the area under cultivation, keep inventing new methods to counteract declining soil fertility, or face agricultural decline owing to degradation of soil fertility or gradual loss of the soil itself.

As agriculture spread north and west, people opened the first clearings in Europe's ancient forest to cultivate small plots for a few years at a time. Ash from burned vegetation fertilized newly cleared fields, helping to maintain initial crop yields until soil fertility declined enough to make it worth the hassle of moving on. The practice of abandoning worn-out fields periodically left fallow land to revegetate, first with grasses, then shrubs, and eventually back to forest. Ground cultivated for a few years then lay fallow for decades as the recolonizing forest gradually revived the soil, allowing clearing and planting again decades later.

Lake sediments, floodplain deposits, and soils record the postglacial evolution of the European landscape. From 7000 to 5500 BC stable environmental conditions left little evidence of human impact. Pollen preserved in lakebeds shows that Neolithic farmers opened clearings in dense forest as agriculture spread north from the Balkans. Cereal pollen shows up in soil profiles and sediment cores about 5500 BC in central Europe. Sediment cores from lakes provide the first incontrovertible evidence of substantial human impact on the central European landscape as massive amounts of charcoal and increased sedimentation—evidence for accelerated soil erosion—coincide with pollen evidence of extensive forest clearing and cereal cultivation around 4300 BC, when postglacial temperatures in Europe were at a maximum.

Farmers had arrived, but Europe was still wild. Lions and hippopotamuses lived along the Thames and Rhine rivers. While scattered bands of people foraged around Europe's lakes, rivers, and coast, a rich soil developed beneath huge oak, elm, and beech trees that stabilized loess-mantled slopes.

Germany's first farmers were drawn to the forest soil developed on silt dropped by glacial winds between the Rhine and Danube rivers. Several centuries later a second wave of related arrivals settled across northern Europe in a band stretching from Russia to France. Soon farmers grew wheat, barley, peas, and lentils on the region's fertile loess. Hunting and gathering thrived outside the loess belt.

Neolithic farmers kept livestock and lived in large longhouses near fields along rivers and streams. Houses were occupied for the several decades that the surrounding fields were kept under continuous cultivation. As isolated longhouses began coalescing into small hamlets, farming spread beyond the loess. More land was cleared and cultivated more continuously. By about 3400 BC hunting for survival was history throughout central Europe.

German soils record periods of agriculturally induced soil erosion from hillsides followed by periods of soil formation lasting roughly five hundred to a thousand years. Soil profiles and alluvial sediments in southern Germany's Black Forest record several periods of rapid erosion associated with increased population. Neolithic artifacts in truncated soil profiles show that initial erosion after the arrival of agriculture about 4000 BC culminated in extensive soil loss by 2000 BC. Declining cereal pollen and a period of soil formation characterized a thousand years of lower population density until renewed erosion in Roman times peaked in the first centuries AD. A second cycle of agricultural decline, soil formation, and forest expansion followed until renewed population growth in the Middle Ages initiated a third, ongoing cycle.

The soils at Frauenberg, a Neolithic site in southeastern Germany, record erosion of nearly the entire soil profile that began with early Bronze Age agriculture. Located on a hill that rises three hundred feet inside a bend in the Danube River, the site's combination of loess soils and a sweeping view of the surrounding country appealed to prehistoric farmers. Remnants of the original soil found in excavations at the site document three distinct periods of occupation corresponding to Bronze Age farming, a Roman fort, and a medieval monastery. Radiocarbon dating of charcoal pulled from soil horizons show that little erosion occurred as soil developed after deglaciation—until Bronze Age farming exposed clay-rich subsoil at the ground surface and eroded nearly the entire loess cover. Subsequent erosion slowed once the less erodible subsoil was exposed. Forest currently blankets the site, which still has limited agricultural potential.

Evidence from soils, floodplains, and lake sediments at sites across Germany shows that human impact has been the dominant influence on the landscape since the last glaciation. Erosion and human occupation occurred in tandem but not in regional pattern as expected for climate-driven events. Just as in ancient Greece and around the Mediterranean, central European cycles of agricultural clearing and erosion associated with population growth gave way to migration, population decline, and renewed soil formation.

Surveys of truncated hillslope soils at more than eight hundred sites along the Rhine River indicate that post-Roman agriculture stripped up to several feet of soil from hillslopes cleared of native forest. Erosion since AD 600 has been about ten times the erosion rate before forest clearing through erosive runoff across bare, plowed fields. Similar soil surveys in Luxembourg report an average of twenty-two inches of lost soil and accelerated soil loss over more than 90 percent of the landscape. Despite the prevalence of Neolithic farming on central Europe's slopes, most of the region's modern agricultural land lies on valley-bottom deposits of reworked soil eroded off surrounding slopes.

Neolithic settlements in southern France are concentrated almost exclusively on limestone plateaus known today for bare white slopes sporting thin, rocky soil and sparse vegetation. When farmers arrived, these uplands were covered by thick brown soil that was far easier to plow than clay-rich valley bottoms. No longer suitable for cultivation, and considered something of a backwater, the limestone plateaus around Montpellier are used primarily for grazing. The harbor at nearby Marseille began filling with sediment soon after Greek colonists founded the city in 600 BC. Sedimentation in the harbor increased thirtyfold after agricultural clearing spread up the steep slopes around the new town.

Early forest clearing in Britain led to extensive soil erosion long before the Roman invasion, as a growing population slowly cleared the forest to plow the slopes. High population density in Roman times exacerbated the loss, in part because better plows worked more of the landscape more often. The population fell dramatically as the empire collapsed, and took almost a thousand years to build back to the same level.

Floodplain sediments along Ripple Brook, a small tributary of the River Severn typical of lowland Britain, record a dramatic increase in deposition rates (and therefore hillslope soil erosion) in the late Bronze Age and early Iron Age. The relative abundance of tree pollen recovered from valley bottom sediments shows that between 2,900 and 2,500 years ago the heavily forested landscape was cleared and intensively farmed. A fivefold increase in floodplain sedimentation speaks to a dramatic increase in hillslope erosion.

Net soil loss averages between three and six inches since woodland clearance in England and Wales. Some watersheds have lost up to eight inches of topsoil. Although much of the loss occurred in the Bronze Age or Roman times, in some places substantial erosion occurred after medieval times. Just two hundred years after Nottinghamshire's famed Sherwood Forest was cleared for agriculture, the original forest soil has been reduced to a layer of thin brown sand over rock. Just as in Lebanon's ancient cedar forest, most of the topsoil is now gone from Robin Hood's woods.

Across the border in Scotland, radiocarbon dating of a sediment core recovered from a small lake west of Aberdeen provides a continuous record of erosion from the surrounding slopes for the past ten thousand years. Sediment deposition rates in the lake, and thus erosion rates on the surrounding slopes, were low for five thousand years under postglacial shrubland and birch forest. Following the arrival of agriculture, pollen from crops and weeds coincide with a threefold increase in the sediment deposition rate. After the Bronze and Iron Ages, erosion decreased dramatically for almost two thousand years as native plants regenerated across a largely abandoned landscape—until erosion accelerated again in the modern era.

Similar cores taken from small lakes in southern Sweden also record the transition from little preagricultural erosion to much higher rates after arrival of the plow. One from Lake Bussjösjö shows that forest stabilized the landscape from 7250 to 750 BC until erosion accelerated following forest clearing. Erosion increased further under intensified agriculture in the sixteenth and seventeenth centuries. A core pulled from Havgårdssjön provides a 5,000-year record of vegetation and erosion. The archaeological record around the lake has no Bronze or Iron Age artifacts. Lakebed sediments piled up four to ten times faster after agricultural settlement began around AD 1100. All across the glaciated terrain of Scandinavia, Scotland, and Ireland, farmers could not make a living until enough ice-free time passed to build soil capable of sustaining cultivation.

Put simply, European prehistory involved the gradual migration of agricultural peoples, followed by accelerated soil erosion, and a subsequent period of low population density before either Roman or modern times. Just as in Greece and Rome, the story of central and western Europe is one of early clearing and farming that caused major erosion before the population declined, and eventually rebounded.

As the Roman Empire crumbled, the center of its civilization shifted north. Abandoning Rome as the capital, Diocletian moved his government to Milan in AD 300. When Theodoric established the Gothic kingdom of Italy on the ruins of the Roman Empire, he chose Verona as his new capital in the north. Even so, many of northern Italy's fields lay fallow for centuries until an eleventh-century program of land reclamation began returning them to cultivation. After several centuries of sustained effort, most of northern Italy's arable land was again under cultivation, supporting prosperous medieval cities that nurtured a renaissance of literature and art.

As northern Italy's population rebounded, intensive land use increased silt loads in the region's rivers enough to attract the attention of Leonardo da Vinci and revive the Roman art of river engineering and flood control. Intensive cultivation on hillside farms spread into the Alps, producing similar results on the Po River as Roman land use had on the Tiber River. Eventually, after eight centuries of renewed cultivation, even northern

Italy's soil faltered. Mussolini's Fascist government spent about half a billion dollars on soil conservation in the 1930s.

Because Rome imported most of its grain from North Africa, Egypt, and the Middle East, it made fewer demands on the soils of the Po Valley, Gaul (France), Britain, and the Germanic provinces. Roman agriculture in its western European provinces was mostly confined to river valleys; for the most part hillslopes that had been farmed in the Bronze Age remained forested until medieval times. It is no coincidence that these northern provinces fed the western European civilization that centuries later rose from the ruins of the Roman Empire.

After the empire collapsed, many Roman fields north and west of the Alps reverted to forest or grass. In the eleventh century, farmers worked less than a fifth of England. With half in pasture and half in crops left fallow every other year, this meant that only about 5 percent of the land was plowed each year. Less than 10 percent of Germany, Holland, and Belgium were plowed annually in the Middle Ages. Even in the most densely populated parts of southern France, no more than 15 percent of the land was cultivated each year.

In early medieval times, townships controlled a given area of land held in common by all villagers. Each household received a share of land to cultivate each season, after which the fields reverted to communal use. The general rule was to plant a crop of wheat, followed by beans and then a fallow season. After the harvest, cattle wandered the fields turning crop stubble into meat, milk, and manure.

Columbia University professor Vladimir Simkhovitch saw the structure of medieval village communities as an adaptation to farming degraded soils. He noted that a similar pattern of land use and ownership characterized many old villages throughout Europe where the land holdings of individual peasants had not been fenced off and enclosed. Barns, stables, and vegetable gardens were always near homesteads, but fields were divided into a patchwork of land belonging to individual farmers. Each farmer might own ten or more parts of three different fields managed collectively for a crop of wheat or rye, then oats, barley, or beans, and finally fallow pasture.

Simkhovitch argued that an inconvenient arrangement in which a farmer had no say in the rotation or type of tillage used on his fields—which could be quite distant from each other—must have been adopted throughout the continent for good reason. He doubted that such arrangements were simply inherited from Roman villas or imposed under feudal-

Figure 10. Miniature from an early-sixteenth-century manuscript of the Middle English poem *God Spede ye Plough* (original held at the British Museum).

ism. Simkhovitch hypothesized that an individual farmer could not keep enough cattle to maintain the fertility of his plot, but a village's livestock could collectively fertilize the commons enough to slow their degradation. Simkhovitch believed that the already degraded state of the land made cooperation the way to survive—a notion contrary to the "tragedy of the commons" in which collective farming was thought to have caused land degradation in the first place.

Simkhovitch argued that by failing to maintain their soil, ancient societies failed themselves. "Go to the ruins of ancient and rich civilizations in Asia Minor, Northern Africa or elsewhere. Look at the unpeopled valleys, at the dead and buried cities. . . . It is but the story of an abandoned farm on a gigantic scale. Depleted of humus by constant cropping, land could no longer reward labor and support life; so the people abandoned it."[1] The introduction of alfalfa and clover into European agriculture helped rebuild soil fertility, Simkhovitch insisted. Noting that there were no hay fields before the sixteenth and seventeenth centuries, he suggested that enclosure of common fields allowed converting enough land to pasture to raise the cattle and sheep needed to manure the land and thereby increase crop yields.

The conventional explanation for the low crop yields of medieval agriculture invoked a lack of enough pasture to supply cultivated land with the manure needed to sustain soil fertility. Until recently, historians generally considered this to reflect ignorance of the value of manure in maintaining soil fertility. It now seems as likely that medieval farmers knew that keeping land in pasture restored soil fertility, but impatience and economics made the required investments unattractive to folks perpetually focused on maximizing this year's harvest.

After centuries when post-Roman agricultural methods and practices limited crop yields, population growth accelerated when an extended run of good weather increased crop yields during medieval times. As the population grew, the clearing of Europe's remaining forests began again in earnest as new heavy plows allowed farmers to work root-clogged lowlands and dense river valley clays. From the eleventh to the thirteenth century, the amount of cultivated land more than doubled throughout western Europe. Agricultural expansion fueled the growth of towns and cities that gradually replaced feudal estates and monasteries as the cornerstone of Western civilization. Europe's best soils had been cleared of forest by about AD 1200. By the close of the thirteenth century, new settlements began plowing marginal lands with poor soils and steep terrain. Expansion of the area of planted fields allowed the population to keep growing. Doubling over a couple of centuries, by AD 1300 Europe's population reached eighty million.

Powerful city-states arose where the most land was plowed under, particularly in and near the fertile lowlands of Belgium and Holland. By the middle of the fourteenth century, farmers were plowing most of western Europe's loess to feed burgeoning societies and their new middle class. Already hemmed in by powerful neighbors, Flemish and Dutch farmers adopted crop rotations similar to those still used today.

The catastrophic European famine of 1315–17 provides a dramatic example of the effect of bad weather on a population near the limit of what its agricultural system could support. Every season of 1315 was wet. Water-logged fields ruined the spring sowing. Crop yields were half of normal and what little hay grew was harvested wet and rotted in barns. Widespread food shortages in early 1316 compelled people to eat the next year's seed crop. When wet weather continued through the summer, the crops failed again, and wheat prices tripled. The poor could not afford food and those with money—even kings—could not always find it to buy. Bands of starving peasants turned to robbery. Some even reportedly resorted to cannibalism in famine-stricken areas.

Malnutrition and starvation began to haunt western Europe. The population of England and Wales had grown slowly but steadily after the Norman invasion until the Black Death of 1348. Major famines added to the toll. The population of England and Wales fell from about four million in the early 1300s to about two million by the early 1400s. Europe's population dropped by a quarter.

After the Black Death depopulated the countryside, landlords competed to retain tenant farmers by granting them lifelong or inheritable rights to the land they worked in exchange for modest rents. As the population rebounded, a final push of agricultural expansion filled out the landscape with farms in the early sixteenth century. Starting in the late 1500s landlords motivated by the promise of getting higher rents from leasing land at inflated rates began enclosing lands formerly grazed in common. Already out of land and surrounded by powerful neighbors, the Dutch started their ambitious campaign to take land from the sea.

John Fitzherbert's 1523 *Book of Surveying,* the first work on agriculture published in English, held that the way to increase the value of a township was to consolidate rights to common fields and pasture into single enclosed tracts next to each farmer's house. Over the next several centuries this idea of reorganizing the commons to give every farmer three acres and a cow evolved into transforming the English countryside into large estates, portions of which could be rented out profitably to tenant farmers. Except for the peasants working the land, most thought that privatizing the commons would injure none, and benefit all by increasing agricultural production.

In the tumultuous sixteenth and seventeenth centuries much of England's agricultural land changed hands in Henry VIII's war against the Catholic Church, the wars of succession, and the English Civil War. The insecurity of land tenure discouraged investing in land improvement. By the second half of the seventeenth century, some argued that England should adopt the Flemish custom of agricultural leases under which the owner would pay a specified sum to the tenant if four impartial persons, two selected by the landlord and two by the tenant, agreed that the soil had been improved at the end of the lease.

As Europe's climate slid from the medieval warm period into the Little Ice Age (which lasted from about AD 1430 to 1850), extended cold periods meant shorter growing seasons, reduced crop yields, and less arable land. Perennially living on the edge, the lower classes were vulnerable to severe food shortages after bad harvests. Governments monitored the price of bread to gauge the potential for social instability.

Desire for land reform among the peasantry, fueled by instability and shortages, would help trigger the Reformation. Land held by the Church had grown over the centuries far beyond the fields cleared by monks, because the Church seldom relinquished land bequeathed by the faithful. Instead, bishops and abbots rented out God's land to poor, land-hungry peasants. By the fifteenth century the Church, which owned as much as four-fifths of the land in some areas, overtook the nobility as Europe's largest landlord. Monarchs and their allies seeking to seize church lands harnessed widespread resentment among tenants. Popular support for the Reformation rested as much on desire for land as the promise of religious freedom.

An increasing demand for crops meant less pasture, little overwinter animal fodder, and not enough manure to sustain soil fertility. As the population kept rising, intensively cultivated land rapidly lost productive capacity—increasing the need to plant more marginal land. Shortage of vacant land to plow helped motivate the rediscovery of Roman agricultural practices such as crop rotations, manuring, and composting.

Renewed curiosity about the natural world also stimulated agricultural experimentation. In the sixteenth century, Bernard Palissy argued that plant ashes made good fertilizer because they consisted of material that the plants had pulled from the soil and could therefore reuse to fuel the growth of new plants. In the early 1600s Belgian philosopher Jan Baptista van Helmont tried to settle the question of whether plants were made of earth, air, fire, or water. He planted a seedling tree in two hundred pounds of soil, protected it from dust, and let it grow for five years, adding only water. Finding that the tree had grown by one hundred and seventy pounds, whereas the soil had lost an insignificant two ounces, van Helmont concluded that the tree had grown from water—the only thing added to the process. Given that the soil had lost but a minuscule fraction of the weight of the tree, he dismissed the potential for earth to have contributed to the tree's growth. I doubt that he ever seriously considered air as a major contributor to the mass of the tree. It took a few more centuries before people discovered carbon dioxide and came to understand photosynthesis.

In the meantime, agricultural "improvers" came to prominence in the seventeenth century once the landscape was fully cultivated. Most of the low hills and shallow valleys of the Netherlands are covered by quartz-rich sand ill suited for agriculture. Supporting a growing population on their naturally poor soils, the Dutch began mixing manure, leaves, and other organic waste into their dirt. Working relatively flat land where erosion was not a problem, over time they built up dark, organic-rich soils to as much as three feet thick. Lacking more land, they made soil. As the Dutch had, the Danes improved their sandy dirt enough to more than double their harvests by adopting crop rotations including legumes and manure. In other words, they readopted key elements of Roman agriculture.

Soil improvement theories spread to England where population growth motivated innovation to increase crop yields. Seventeenth-century agriculturalists broadened the range of fodder crops, developed more complex crop rotations, used legumes to improve soil fertility, and used more manure to maintain soil fertility. In addition, introduction of the Flemish practice of growing clover and turnips as ground cover and winter fodder changed the ratio of animals to land, increasing the availability of manure. Improvers promoted clover as a way to rejuvenate fields and regain high crop yields: clover increased soil nitrogen directly through the action of nitrogen-fixing bacteria in nodules on the plant roots and, as feed for cattle, also produced manure.

Despite cold winters, wet summers, and shorter growing seasons, English agriculture increased its yields per acre from 1550 to 1700, in what has been called the "yeoman's agricultural revolution." At the start of the seventeenth century, between a third and half of English agricultural land was held by yeomen—small freehold farmers and those with long-term leases. In the early 1600s farmers obsessed with fertilization began plowing into their fields lime, dung, and almost any other organic waste that could be obtained. Farmers also began shifting away from fixed grain lands and pasture and began planting fields for three or four years, and then putting them to pasture for four or five years before plowing them up once again. This new practice of "convertible husbandry" resulted in much higher crop yields, making it attractive to plow up pastures formerly held in common.

The new breed of land improvers also pioneered systems to drain and farm wetlands. They experimented with plow design and with ways to improve soil fertility. Upper-class landowners advocated enclosing pasturelands and growing fodder crops (especially turnips) to provide winter feed for cattle and increase the supply of manure. Adopting the premise that communal land use degraded the land, an idea now called the "tragedy of the commons," agricultural improvers argued that enclosing the commons into large estates was necessary to increase agricultural output. A Parliament of property owners and lawyers passed laws to fence off fields that had been worked in common for centuries. Land enclosures increased crop yields and created tremendous wealth for large landowners, but the peasants thus fenced out—whose parents ate meat, cheese, and vegetables raised by their own efforts—were reduced to a diet of bread and potatoes.

Soil husbandry began to be seen as the key to productive, profitable farming. Gervase Markham, one of the first agricultural writers to write in English instead of Latin, described soils as various mixtures of clay, sand, and gravel. What made good soil depended on the local climate, the character and condition of the soil, and the local plants (crops). "Simple Clays, Sands, or Gravels together; may be all good, and all fit to bring forth increase, or all . . . barren." Understanding the soil was the key to understanding what would grow best, and essential to keeping a farm productive. "Thus having a true knowledge of the Nature and Condition of your ground, . . . it may not only be purged and clensed . . . but also so much bettered and refined."[2]

Prescribing steps to improve British farms, Markham recommended using the right type of plow for the ground. He advised mixing river sand and crushed burned limestone into the soil, to be followed by the best manure to be had—preferably ox, cow, or horse dung. In describing procedures for improving barren soils, Markham advocated growing wheat or rye for two years in a field, and then letting sheep graze and manure it for a year. After the sheep, several crops of barley were to be followed in the seventh year by peas or beans, and then several more years as pasture. After this cycle the ground would be much improved for growing grain. The key to sustaining soil fertility was to alternate livestock and crops on the same piece of ground.

Equally important, although it received less attention, was preventing erosion of the soil itself. Markham advised plowing carefully to avoid collecting water into erosive gullies. Good soil was the key to a good farm, and keeping soil on the farm required special effort even on England's gentle rolling hills.

Almost half a century later, on April 29, 1675, John Evelyn presented a "Discourse on Earth, Mould and Soil" to England's Royal Society for the Improvement of Natural Knowledge. In addressing what he feared could be considered a topic unworthy of the assembled luminaries he invited the society's fellows to descend from contemplating the origin of heavenly bodies and focus instead on the ground beneath their feet. He implored them to consider both how soil formed and how the nation's long-term prosperity depended on improving the kingdom's dirt.

Evelyn described how distinct layers of topsoil and subsoil developed from the underlying rock. "The most beneficial sort of Mould or Earth, appearing on the surface . . . is the natural (as I beg leave to call it) underturf-Earth and the rest which commonly succeeds it, in Strata's or layers, 'till we arrive to the barren, and impenetrable Rock." Of the eight or nine basic types of soil, the best was the rich topsoil where mineral soil mixed with vegetation.

> I begin with what commonly first presents it self under the removed Turf, and which, for having never been violated by the Spade, or received any foreign mixture, we will call the Virgin Earth; . . . we find it lying about a foot deep, more or less, in our Fields, before you come to any manifest alteration of colour or perfection. This surface-Mold is the best, and sweetest, being enriched with all that the Air, Dews, Showers, and Celestial Influences can contribute to it.

The ideal topsoil was a rich mixture of mineral and organic matter introduced by "the perpetual and successive rotting of the Grass, Plants, Leaves, Branches, [and] Moss . . . growing upon it."[3]

Regaling his audience with the works of Roman agriculturalists, Evelyn described how to improve soil with manure, cover crops, and crop rotations. Like the Romans, Evelyn used odor, taste (sweet or bitter), touch (slippery or gritty), and sight (color) to evaluate a soil. He described different types of manure and their effects on soil fertility, as well as the virtues of growing legumes to improve the soil.

Echoing Xenophon, Evelyn held that to know the soil was to know what to plant. One could read what would grow best by observing what grew naturally on a site. "Plants we know, are nourished by things of like affinity with the constitution of the Soil which produces them; and therefore 'tis of singular importance, to be well read in the Alphabet of Earths and Composts." Because soil thickened as organic material supplied from above mixed with the rotting rocks below, sustaining good harvests required maintaining the organic-rich topsoil ideal for crops. Mineral subsoil was less productive, but Evelyn believed that nitrous salts could resuscitate even the most exhausted land. "I firmly believe, that were Salt Peter . . . to be obtain'd in Plenty, we should need but little other Composts to meliorate our Ground."[4] Well ahead of his time, Evelyn anticipated the value of chemical fertilizers for propping up—and pumping up—agricultural production.

By the start of the eighteenth century, improving farmland was seen as possible only through enclosing under private ownership enough pasture to keep livestock capable of fertilizing the plowed fields. Simply letting the family cow poop on the commons would not do. The need for manure imposed an inherent scale to productive farms. Too small a farm was a

The Whole ART
O F
HUSBANDRY;
Or, The Way of
Managing and *Improving*
O F
LAND.

B E I N G

A full COLLECTION of what hath been
Writ, either by ancient or modern Authors: With
many Additions of new Experiments and Improvements not treated of by any others.

A S   A L S O,

An ACCOUNT of the particular Sorts of
Husbandry ufed in feveral *Counties*; with Propofals
for its farther Improvement.

To which is added,

The Country-man's Kalendar, what he is to do
every Month in the Year.

By *J. Mortimer*, Efq; *F. R. S.*

The Second Edition, Corrected.

L O N D O N,

Printed by *J. H.* for *H. Mortlock* at the *Phœnix*, and
J. Robinson at the *Golden Lion* in St. *Paul's* Church-
Yard, M DCC VIII.

Figure 11. Title page to *The Whole Art of Husbandry,* published in 1708.

recipe for degrading soil fertility through continuous cropping. Although very large farms turned out to mine the soil itself, this was not yet apparent—and Roman experience in this regard was long forgotten. To the individual farmer, enclosure was seen as the way to ensure a return on investing to improve soil fertility from well-manured ground.

Agricultural writers maintained that the key to good crop yields was to keep an adequate supply of manure on hand—to keep the right ratio of pasture to field on each farm, or estate as the case increasingly became. "The Arable-land must be proportioned to the quantity of Dung that is raised in the Pasture, because proper Manure is the chief Advantage of Arable-ground."[5] The key to increasing agricultural productivity was seen to lie in bringing stock raising and cereal production into proximity and returning manure to the fields.

Still, not all land was the same; improvements needed to be tailored to the nature of the soil. British farmland consisted of three basic types: uplands lying high enough not to flood, lowlands along rivers and wetlands, and land susceptible to inundation by the sea. These lands had different vulnerabilities.

On hillslopes, the thin layer of a foot or so of topsoil was essential to good farming. Such lands were naturally prone to erosion and vulnerable to poor farming practices. On lowlands, the soil was replenished by upland erosion that produced fine deposits downslope. "As to Lands lying near Rivers, the great Improvement of them is their over-flowing, which brings the Soil of the Uplands upon them, so as that they need no other mending though constantly mowed."[6]

Working land too hard for too long would reduce soil fertility. Sloping land was particularly vulnerable. "Where Lands lie upon the sides of Hills . . . great care must be taken not to plow them out of heart."[7] Recognizing such connections, most landlords obliged their tenants to fallow fields every third year, and every other year if manure was unavailable. Reviving worn-out fields proved highly profitable—when enough land was enclosed. Under the banner of agricultural improvement, Parliament repeatedly authorized land enclosures that created large estates at the expense of common land, enriching the landed gentry and turning peasants into paupers.

English farmers gradually increased per-acre grain yields to well above medieval crop yields of twice the seeded amount, which were no greater than early Egyptian crop yields. Traditionally, historians attributed increased yields between the Middle Ages and the Industrial Revolution to the introduction of clover and other nitrogen-fixing plants into crop rotations in the eighteenth and early nineteenth centuries. Crop yields at the start of the eighteenth century were not all that much greater than medieval levels, implying that increased agricultural production came largely from expanding the area cultivated rather than improved agricultural methods. Wheat yields had risen by just a bushel and a half over medieval yields of ten to twelve bushels per acre. Yet by 1810 yields had almost doubled. By 1860 they had reached twenty-five to twenty-eight bushels an acre.

Increasing labor needed to harvest an acre of crops implies that crop yields rose over time. The number of person days required to harvest an acre of wheat increased from about two around 1600 to two and a half by the early 1700s, and then to just over three in 1860. Overall crop yields increased by two and a half times in the six hundred years from 1200 to 1800. So despite increasing yields, the tenfold population increase primarily reflected expansion of the area under cultivation.

During the same period about a quarter of England's cultivated land was transformed from open, common fields to fenced estates. By the end of the eighteenth century, common fields had almost disappeared from the English landscape. Loss of the common lands meant the difference between independence and destitution for rural households that had always kept a cow on the commons. Dispossessed, landless peasants with no work depended on public relief for food. Seeing the economic effects of the transformation of the English countryside, Board of Agriculture secretary Arthur Young came to see land enclosure as a dangerous trend destroying rural self-sufficiency. But enclosing and privatizing the last vestiges of communal property conveniently pushed a new class of landless peasants to seek jobs just as laborers were needed in Britain's industrializing cities.

By the early nineteenth century, British farms had developed into a mixed system of fields and pastures. A roughly equal emphasis on cultivation and animal husbandry provided for constant enrichment of the soil with large quantities of manure, and cover crops of clover and legumes.

English population growth mirrored increases in agricultural production from after the Black Death to the Industrial Revolution. Between 1750 and 1850, England's cereal production and population both doubled. Did a growing human population drive up demand for agricultural products? Or did increased agricultural production enable faster population growth? Regardless of how we view the causality, the two rose in tandem.

Nonetheless, as the population grew, the European diet declined. With almost all of the available land in cultivation, Europeans increasingly survived on vegetables, gruel, and bread. Without surplus grain to feed animals through the winter, and later without access to the commons to graze cattle, eating meat became an upper-class privilege. An anonymous pamphlet published in London in 1688 attributed massive unemployment to Europe's being "too full of people" and advised wholesale emigration to America. At the start of the nineteenth century, most Europeans survived on 2,000 calories a day or less, about the average for modern India and below the average for Latin America and North Africa. European peasants toiling in their fields ate less than Kalahari Desert bushmen who worked just three days a week.

Despite increased agricultural production, food prices rose dramatically in both England and France during the sixteenth and seventeenth centuries. Persistent famine between 1690 and 1710 stalked a population larger than could be reliably fed. While enlightened Europe lived on the edge of starvation, Britain largely escaped the peasant unrest that sparked the French Revolution by importing lots of food from Ireland.

Real hunger, as much as the hunger for empire or religious freedom, helped launch Europe toward the New World. Beginning with Spain, the thickly settled and most continuously cultivated parts of western Europe most aggressively colonized the New World. Before the Romans, the Phoenicians and Greeks had settled Spain's eastern coast, but Iberian agriculture remained primitive until aggressive Roman cultivation. The Moors introduced intensive irrigation to Spain a few centuries after the fall of Rome. More than five hundred years of Moorish agriculture further degraded Spanish soils. By the fifteenth century, the fertile soils of the New World looked good to anyone working Spain's eroded and exhausted soil. Within a few generations, Spanish and Portuguese farmers replaced gold-seeking conquistadors as the primary emigrants to Central and South America.

By contrast, it took more than a century after Columbus for northern European farmers to begin heading west for religious and political freedom—and tillable land. English and French peasants were still clearing and improving land in their own countries. German peasants were busy plowing up newly acquired church land. Germany did not even begin to establish overseas colonies until the 1850s. The northern European rush to America did not kick into full gear until the late nineteenth century. Relatively few people from northwestern Europe migrated to America while there was still fertile land at home.

As continental Europe filled in with farms, peasants moving up into the hills set the stage for crisis once eroding slopes could no longer support a

Figure 12. Mid-eighteenth-century agricultural landscape (Diderot's *Encyclopédie,* Paris, 1751–80).

hungry population. When eighteenth-century farmers began clearing steep lands bordering the French Alps, they triggered landslides that carried off soils and buried valley bottom fields under sand and gravel. By the late eighteenth century, the disastrous effects of soil erosion following deforestation of steep lands had depopulated portions of the Alps. Nineteenth-century geographer Jean-Jacques-Élisée Reclus estimated that the

French Alps lost a third to more than half their cultivated ground to erosion between the time Columbus discovered America and the French Revolution. By then people crowding into cities in search of work could neither grow nor pay for food.

A decade of persistent hunger laid the groundwork for revolution as the homeless population of Paris tripled. According to the bishop of Chartres, conditions were no better in the countryside, where "men were eating grass like sheep, and dying like flies." Revolutionary fervor fed on long lines at bakeries selling bitter bread full of clay at exorbitant prices. Anger over the price of the little available for sale and the belief that food was being withheld from the market spurred on the mobs during key episodes of the French Revolution.

Dissolution of the nobility's large estates freed peasants to grab still forested uplands. Clearing steep slopes triggered debris torrents that scoured uplands and buried floodplain fields under sand and gravel. Large areas of upper Provence were virtually abandoned. Between 1842 and 1852 the area of cultivated land in the lower Alps fell by a quarter from the ravages of landslides and soil erosion.

French highway engineer Alexandre Surell worked on devising responses to the landslides in the Upper Alps (Hautes-Alpes) in the early 1840s. He noted the disastrous consequences that followed when cultivation pushed into the mountains. Torrents cascading off denuded slopes buried fields, villages, and their inhabitants. Everywhere the forests had been cut there were landslides; there were no landslides where the forest remained. Connecting the dots, Surell concluded that trees held soil on steep slopes. "When the trees became established upon the soil, their roots consolidate and hold it by a thousand fibres; their branches protect the soil like a tent against the shock of sudden storms."8

Recognizing the connections between deforestation and the destructive torrents, Surell advocated an aggressive program of reforestation as the way to a secure livelihood for the region's residents. Plowing steep land was an inherently short-term proposition. "In the first few years following a clearing made in the mountains, excellent crops are produced because of the humus coat the forest has left. But this precious compost, as mobile as it is fecund, lingers not for long upon the slopes; a few sudden showers dissipate it; the bare soil quickly comes to light and disappears in its turn."9 Measures to protect the forest and the soil were often unsuccessful because it was more immediately profitable to clear and plant, even though deforested slopes could not be farmed for long.

While Surell fretted about how to restore upland forests, George Perkins Marsh toured France during his service as American ambassador in Italy. Witnessing the long-term effects of forest clearing on both steep land and valley fields, Marsh saw that bare, eroded mountain slopes unfit for habitation no longer absorbed rainfall but rapidly shed runoff that picked up sediment and dumped it on valley fields.

An observant tourist, Marsh feared the New World was repeating Old World mistakes.

> The historical evidence is conclusive as to the destructive changes occasioned by the agency of man upon the flanks of the Alps, the Apennines, the Pyrenees, and other mountain ranges in Central and Southern Europe, and the progress of physical deterioration has been so rapid that, in some localities, a single generation has witnessed the beginning and the end of the melancholy revolution. . . . It is certain that a desolation, like that which has overwhelmed many once beautiful and fertile regions of Europe, awaits an important part of the territory of the United States, and of other comparatively new countries over which European civilization is now extending its sway.[10]

Marsh compared what he saw in Europe to New York State, where the upper Hudson River was filling with sediment as farmers plowed up the forest. He held that gentle slopes in areas where rainfall was evenly distributed throughout the seasons could be reasonably farmed on a permanent basis. Ireland, England, and the vast Mississippi basin fit this definition. In contrast, steep terrain could not be plowed for long without triggering severe erosion, especially in regions with torrential rains or parching droughts.

French deforestation peaked in the early 1800s. In 1860 the marquis de Mirabeau estimated that half of France's forest had been cleared in the previous century. Inspector of Forests Jonsse de Fontanière echoed Surell's stark assessment of the prospects of the High Alps. "The cultivators of the land . . . will be compelled . . . to abandon the places which were inhabited by their forefathers; and this solely in consequence of the destruction of the soil, which, after having supported so many generations, is giving place little by little, to sterile rocks."[11]

French authorities began passing laws to protect and restore public and private woodlands in 1859. Clearing of European forests accelerated briefly, though, when twenty-eight thousand walnut trees were cut to supply European manufacturers with gunstocks during the American Civil War.

Despite such profiteering, by 1868 almost two hundred thousand acres of the High Alps had been replanted with trees or restored to meadow.

Touring southern France before the Second World War, Walter Lowdermilk found intensive farming practiced on both steep slopes and valley floors. Some farmers maintained hillslope terraces like those built by the ancient Phoenicians. Lowdermilk marveled over how in eastern France, where terracing was uncommon, farmers would collect soil from the lowest furrow on a field, load it into a cart, haul it back up the slope, and dump it into the uppermost furrow. Centuries ago when this practice began, peasant farmers knew that they had upset the balance between soil production and erosion, and that people living on the land would inherit the consequences. They probably did not appreciate how far they were ahead of Europe's gentlemen scientists in understanding the nature of soils.

At the May 5, 1887, meeting of the Edinburgh Geological Society, vice president James Melvin read from an unpublished manuscript by James Hutton, the Scottish founder of modern geology. The rediscovered work revealed the formative geologic insights Hutton had gained from farming the land, observing and thinking about relationships among vegetation, soil, and the underlying rocks. In particular, Melvin appreciated the parallels between Hutton's century-old musings and Darwin's newly published book on worms.

Hutton saw soil as the source of all life where worms mix dead animals with fallen leaves and mineral soil to build fertility. He thought that hillslope soils came from the underlying rock, whereas valley bottom soils developed on dirt reworked from somewhere upstream. Soil was a mix of broken rock from below and organic matter from above, producing dirt unique to each pairing of rocks and plant communities. Forests generally produced fine soils. "[A forest] maintains a multitude of animals which die and are returned to the soil; secondly, it sheds an annual crop of leaves, which contribute in some measure to the fertility of the soil; and lastly, the soil thus enriched with animal and vegetable bodies feeds the worms . . . which penetrate the soil, and introduce fertility as they multiply."[12] Anticipating Darwin in recognizing the role of worms in maintaining soil fertility, Hutton also understood the role of vegetation in establishing soil characteristics. The visionary geologist saw soil as the living bridge between rock and life maintained by returning organic matter to the soil.

At the close of the eighteenth century—long before Melvin rediscovered Hutton's lost manuscript—Hutton argued with Swiss émigré Jean André de Luc over the role of erosion in shaping landscapes. De Luc held that ero-

Figure 13. French farmers loading soil from their lowest furrow into a cart to be hauled back uphill in the late 1930s (Lowdermilk 1953, 22, fig. 12).

sion stopped once vegetation covered the land, freezing the landscape in time. At issue was whether topography was the ultimate fossil, left over from Noah's flood. Hutton questioned de Luc's view, pointing to the turbid waters of flooding rivers as evidence of erosion endlessly working to lower mountains. "Look at the rivers in a flood;—if these run clear, this philosopher [de Luc] has reasoned right, and I have lost my argument. Our clearest streams run muddy in a flood. The great causes, therefore, for the degradation of mountains never stop as long as there is water to run; although as the heights of mountains diminish, the progress of their diminution may be more and more retarded."[13] In other words, steeper slopes eroded faster, but all land eroded.

A few years later Hutton's disciple, geologist and mathematician John Playfair, described how weathering created new soil at about the rate that erosion removed it. He saw topography as the product of an ongoing war between water and rock. "Water appears as the most active enemy of hard and solid bodies; and, in every state, from transparent vapour to solid ice, from the smallest rill to the greatest river, it attacks whatever has emerged above the level of the sea, and labours incessantly to restore it to the deep."[14]

Adopting Hutton's radical concept of geologic time, Playfair saw how erosion worked gradually to destroy land that dared rise above sea level. Yet the land remained covered by soil despite this eternal battle.

> The soil, therefore, is augmented from other causes, . . . and this augmentation evidently can proceed from nothing but the constant and slow disintegration of the rocks. In the permanence, therefore, of a coat of vegetable mould on the surface of the earth, we have a demonstrative proof of the continual destruction of the rocks; and cannot but admire the skill, with which the powers of the many chemical and mechanical agents employed in this complicated work, are so adjusted, as to make the supply and the waste of the soil exactly equal to one another.[15]

The soil maintained a uniform thickness over time even as erosion continuously reshaped the land.

About the time Hutton and Playfair were trying to convince Europe's learned societies of the dynamic nature of soil over geologic time, parallel arguments about the controls on the size and stability of human populations were brewing. Europeans began questioning the proposition that greater population led to greater prosperity. On an increasingly crowded continent, limits to human population growth were becoming less abstract.

The Reverend Thomas Malthus infamously proposed that a boom-and-bust cycle characterizes human populations in his 1798 *Essay on the Principle of Population*. A professor of political economy at Haileybury College, Malthus argued that exponentially growing populations increase faster than their food supply. He held that population growth locks humanity in an endless cycle in which population outstrips the capacity of the land to feed people. Famine and disease then restore the balance. British economist David Ricardo modified Malthus's ideas to argue that populations rise until they are in equilibrium with food production, settling at a level governed by the amount of available land and the technology of the day. Others like the marquis de Condorcet argued that necessity motivates innovation, and that agriculture could keep up with population growth through technological advances.

Malthus's provocative essay overlooked how innovation can increase crop yields and how greater food production leads to even more mouths to feed. These shortcomings led many to discredit Malthus because he treated food production and food demand as independent factors. He also neglected to consider the time required for agriculturally accelerated erosion to strip topsoil off a landscape or for intensive cultivation to deplete soil fertility. Although his views seemed increasingly naive as England's population kept growing, political interests seeking to rationalize exploitation of Europe's new working class embraced them.

Malthus's ideas challenged prevalent views of human impact on nature in general and on the soil in particular. In *Political Justice,* published five years before Malthus's essay, William Godwin captured the fashionable view of the inevitable progress of human dominion over nature. "Three-fourths of the habitable globe are now uncultivated. The improvements to be made in cultivation, and the augmentations the earth is capable of receiving in the article of productiveness, cannot, as yet, be reduced to any limits of calculation. Myriads of centuries of still increasing population may pass away, and the earth be yet found sufficient for the support of its inhabitants."[16] In Godwin's view, scientific progress promised endless prosperity and ongoing advances in material well-being. The basic perspectives of Malthusian pessimism and Godwinian optimism still frame debates about the relationships between human populations, agricultural technology, and political systems.

Published early in the Industrial Revolution, Malthus's ideas were adopted by those wanting to explain poverty as the fault of the poor themselves, rather than an undesirable side effect of land enclosure and industrial development. Taken at face value, Malthus's ideas absolved those at the top of the economic ladder from responsibility for those at the bottom. In contrast, Godwin's ideas of material progress became associated with the movement to abolish private property rights. Naturally, Malthus would have more appeal for a Parliament of wealthy landowners.

As intellectuals debated the earth's capacity to provide sustenance, the working classes continued to live on the verge of starvation. Vulnerability to bad harvests continued well into the nineteenth century as European agriculture could barely keep up with the rapidly growing cities. High grain prices during the Napoleonic Wars further accelerated land enclosures across Britain. Then in 1815, after the eruption of Indonesia's Tomboro volcano, the coldest summer on record produced catastrophic crop failures. Food riots in England and France spread across the continent when hungry workers faced skyrocketing bread prices. The price of a loaf of bread remained a central point of working-class protest as the discontent of the urban poor bred radicals and revolutionaries.

A potato blight that arrived from America in 1844–45 showed just how insecure food production had become. When *Phytophthora infestans* wiped out the Irish potato harvest in the summer of 1845 and the next year's crop failed too, it left the poor—who could not afford to buy food at market rates from the indifferent British government—with literally nothing to eat. Completely dependent on potatoes, the Irish population crashed. About a million people died from starvation or associated diseases. Another million emigrated during the famine. Three million more left the country over the next fifty years, many bound for America. By 1900 the population of Ireland was a little more than half of what it had been in the 1840s. Why had the Irish become so dependent on a single crop, particularly one introduced from South America only a century before?

At first glance the answer appears to support Malthus. Between 1500 and 1846 the Irish population increased tenfold to eight and a half million. As the population grew, the average land holding dwindled to about 0.2 hectares (half an acre), enough to feed a family only by growing potatoes. By 1840 half the population ate little besides potatoes. More than a century of intensive potato cultivation on nearly all the available land had reduced the Irish to living on the verge of starvation in good years. But a closer look at this story reveals more than a simple tale of population outpacing the ability to grow potatoes.

The potato grew in importance as a staple while Irish agriculture increasingly exported everything else to Britain and its Caribbean colonies. In 1649 Oliver Cromwell had led an invasion to carve Ireland into plantations to pay off with land the speculators who bankrolled Parliament in the English Civil War. Ireland's new landlords saw lucrative opportunities provisioning Caribbean sugar and tobacco plantations. Later, increasing demand for food in Britain's industrializing cities directed Irish exports to closer markets. In 1760 hardly any Irish beef went to Britain. By 1800 four out of five Irish cows sent to market ended up on British tables. The growth of Britain's urban population created substantial demand for food Irish landlords were happy to supply. Even after the official union of Ireland and England in 1801, Ireland was run as an agricultural colony.

The potato increasingly fed rural Ireland as land was diverted to raise exports. In order to devote the best land to commercial crops, landlords pushed peasants onto marginal lands where they could grow little other than potatoes. Adam Smith advocated the potato as a means to improve landlords' profits in *The Wealth of Nations* because tenants could survive on smaller plots if they grew nothing but potatoes. By 1805 the Irish ate little meat. With most of the country's beef, pork, and produce shipped off to Britain, the poor had nothing to eat when the potato crop failed.

There was no relief effort during the famine. On the contrary, Irish exports to England increased. The British Army helped enforce contracts as landlords shipped almost half a million Irish pigs to England at the peak of the famine in 1846. This policy of expedience was not unusual. More food was available during many European famines than was accessible to peasants who had no backup when their crops failed. Poor subsistence farmers could not buy food on the open market. As the ranks of the urban poor grew, they too could not afford food at the higher prices famines produced. And without land they could not feed themselves. Food riots swept across Europe in 1848 in the wake of the potato blight and a poor grain harvest on the continent.

Agricultural economics began to shape radical thought. In the early 1840s, before he met Karl Marx, Friedrich Engels took issue with Malthus and argued that labor and science increased as fast as population and therefore agricultural innovation could keep pace with a growing population. Marx, by contrast, saw commercialized agriculture as degrading to both society and the soil. "All progress in capitalistic agriculture is a progress in the art, not only of robbing the worker, but of robbing the soil; all progress in increasing the fertility of the soil for a given time is a progress towards ruining the more long-lasting sources of that fertility."[17] (Ironically, in the decade before the 1917 Russian Revolution, Czar Nicholas II passed land reforms that began giving peasants title to their land. Unlike the urban poor who rallied to Lenin's promise of "bread, peace, and land," rural peasants were slow to embrace the revolution Marx anticipated they would lead.)

Governments continued to export grain during famines well into the twentieth century. Soviet peasants starved in the 1930s when the central government appropriated their harvest to feed the cities and sell to foreign markets for cash to fund industrialization. In most famines, social institutions or food distribution inequities cause as much hunger as absolute shortages of food.

The initial response to rising population in postmedieval Europe was to bring progressively more marginal land into agricultural production. Yields may have been lower than from traditional farmlands, but the food produced from these lands helped sustain population growth. Starting in the eighteenth century, European powers harnessed the agricultural potential of their colonies around the world to provide cheap imported food. European agricultural self-reliance ended when imports shifted from luxuries such as sugar, coffee, and tea to basic foodstuffs like grains, meat, and dairy products. By the end of the nineteenth century, many European nations depended on imported food to feed their populations.

As Western empires spread around the globe, colonial economics displaced locally adapted agricultural systems. Typically, introduction of European methods replaced a diversity of crops with a focus on export crops like coffee, sugar, bananas, tobacco, or tea. In many regions, sustained cultivation of a single crop rapidly reduced soil fertility. In addition, northern European farming methods developed for flat-lying fields shielded under snow in winter and watered by gentle summer rains led to severe erosion on steep slopes subject to intense tropical rainfall.

Europe solved its perennial hunger problem by importing food and exporting people. About fifty million people left Europe during the great wave of emigration between 1820 and 1930; many European peoples now have more descendants in former colonies than live in the motherlands. Colonial economics and policies that favored plantation agriculture unofficially encouraged soil degradation and perpetual hunger for fresh land. Paradoxically, the drive to establish colonies was itself driven by European land hunger fueled by degradation of upland regions and enclosure of communal farmland into large estates.

Europeans emerged from under the cloud of malnutrition and constant threat of starvation because their colonial empires produced lots of cheap food. Europeans outsourced food production as they built industrial economies. Between 1875 and 1885, a million acres of English wheat fields were converted to other uses. With a growing industrial economy and a shrinking agricultural land base, Britain increasingly ate imports. By 1900 Britain imported four-fifths of its grain, three-quarters of its dairy products, and almost half its meat. Imported food pouring into Europe mined soil fertility on distant continents to further the growth of industrializing economies.

After Europe's colonial empires dissolved at the end of the Second World War, Josué de Castro, chairman of the executive council of the United Nations Food and Agriculture Organization, argued that hunger not only prepared the ground for history's great epidemics but had been one of the most common causes of war throughout history. He viewed the success of the Chinese Revolution as driven by the strong desire for land reform among tenant farmers forced to surrender half their harvest from microscopic fields to owners of huge estates. Mao Ze-dong's strongest ally was the fear of famine. The chairman's most fervent partisans were the fifty million peasants he promised land.

Agitation for land reform in the third world colored the postcolonial geopolitical landscape of the twentieth century. In particular, subsistence farmers in newly independent countries wanted access to the large land holdings used to grow export crops. Since then, however, land reform has been resisted by Western governments and former colonies, who instead stressed increasing agricultural output through technological means. Generally, this meant favoring large-scale production of export crops over subsistence farming. Sometimes it meant changing a government.

In June 1954 a U.S.-backed coup overthrew the president of Guatemala. Elected in 1952 with 63 percent of the vote, Jacobo Árbenz had formed a coalition government that included four Communists in the fifty-six-member Chamber of Deputies. An alarmed United Fruit Company, which held long-term leases to much of the coastal lowlands, launched a propaganda campaign pushing the view that the new Guatemalan government was under Russian control. It's unlikely that the few Communist party members in the government had that much clout; United Fruit's real fear was land reform.

In the late nineteenth century, the Guatemalan government had appropriated communal Indian lands to facilitate the spread of commercial coffee plantations throughout the highlands. At the same time, U.S. banana companies began acquiring extensive lowland tracts and building railways to ship produce to the coast. Export plantations rapidly appropriated the most fertile land and the indigenous population was increasingly pushed into cultivating steep lands. By the 1950s, many peasant families had little or no land even though companies like United Fruit cultivated less than a fifth of their vast holdings.

Soon after coming to power, Árbenz sought to expropriate uncultivated land from large plantations and promote subsistence farming by giving both land and credit to peasant farmers. Contrary to United Fruit's claims, Árbenz did not seek to abolish private property. However, he did want to redistribute more than 100,000 hectares of company-leased land to small farmers and promote microcapitalism. Unfortunately for Árbenz, U.S. Secretary of State John Foster Dulles had personally drafted the banana company's generous ninety-nine-year lease in 1936. With Dulles on United Fruit's side, even the pretense of Communist influence was enough to motivate a CIA-engineered coup in the opening years of the cold war.

Subsequent foreign investment opened more land for cash crops and cattle. International aid and loans from development banks promoted large projects focused on export markets. Between 1956 and 1980, large-scale monoculture projects received four-fifths of all agricultural credit. Land devoted to cotton and grazing grew more than twentyfold. Land planted in sugar quadrupled. Coffee plantations grew by more than half. Forced from the most fertile land, Guatemalan peasants were pushed up hillsides and into the jungle. Four decades after the 1954 coup, fewer than two out of every hundred landowners controlled two-thirds of Guatemalan farmland. As the size of agricultural plantations increased, the average farm size fell to under a hectare, less than needed to support a family.

This was the story of Ireland all over again, with a Latin American twist—Guatemala is a steep country in the rain-drenched tropics. But like Irish meat, Guatemalan coffee is sold elsewhere. And like its coffee, Guatemala's soil is also leaving as adoption of European agricultural methods to tropical hillslopes ensures a legacy of major erosion. The combination of cash crop monoculture and intensive subsistence farming on inherently marginal lands increased soil erosion in Guatemala dramatically, sometimes enough to be obvious to even the casual observer.

In the last week of October 1998, Hurricane Mitch dumped a year's worth of rain onto Central America. Landslides and floods killed more than ten thousand people, left three million displaced or homeless, and caused more than $5 billion in damage to the region's agricultural economy. Despite all the rain, the disaster was not entirely natural.

Mitch was not the first storm to dump that much rain on Central America, but it was the first to fall on the region's steep slopes after the rainforest had been converted into open fields. As the population tripled after the Second World War, unbroken forest surrounding a few cleared fields was replaced by continuously farmed fields. Now, most of the four-fifths of the rural population farm tiny plots on sloping terrain practicing a small-scale version of conventional agriculture. While accelerated erosion from farming Central America's steep slopes has long been recognized as a problem, Hurricane Mitch ended any uncertainty as to its importance.

After the storm, a few relatively undamaged farms stood out like islands in a sea of devastation. When reconnaissance surveys suggested that farms practicing alternative agriculture better survived the hurricane than did conventional farms, a coalition of forty nongovernmental agencies started an intensive study of more than eighteen hundred farms in Guatemala, Honduras, and Nicaragua. Pairing otherwise comparable farms that practiced conventional and so-called sustainable agricultural practices, teams inspected each farm for soil condition, evidence of soil erosion, and crop losses. Across the region, farms operated with sustainable methods such as polyculture, hillside terracing, and biological pest control had two to three times less soil erosion and crop damage than conventional farms under chemical-intensive monoculture. Gullies were less pronounced and landslides were two to three times less abundant on sustainable farms than on conventional farms. Sustainable farms had less economic damage as well. Perhaps the most telling result of the study was that more than nine out of ten of the conventional farmers whose farms were inspected expressed a desire to adopt their neighbors' more resilient practices.

Central America was but one of many regions where the growth of large, export-oriented plantations after the Second World War turned former colonies into agricultural colonies serving global markets. Commercial monocultures also displaced subsistence farmers into marginal lands across Asia, Africa, and South America. In the new global economy, former political colonies continued to serve the interests of wealthier nations—only now trading soil for cash. But this is not all that new: the United States was in the same position before its own revolution.

SIX

Westward Hoe

Since the achievement of our independence, he is
the greatest Patriot, who stops the most gullies.

PATRICK HENRY

SEVERAL YEARS AGO, ON A BREAKNECK research trip down rough dirt roads through a recently deforested part of the lower Amazon, I saw how topsoil loss could cripple a region's economy and impoverish its people. I was there to study caves created over a hundred million years as water slowly dissolved iron-rich rocks that lay beneath soils resembling weathered frying pans. Walking through an iron cave impressed upon my imagination how long it must have taken for dripping water to carve them. Just as striking on this trip were the signs of catastrophic soil loss after forest clearing. Yet what really amazed me was how this human and ecological catastrophe-in-the-making did not change people's behavior, and how the modern story of the lower Amazon paralleled the colonial history of the United States.

Standing on the edge of the Carajás Plateau, I straddled the skeletal remains of an ancient landscape and another still being born. Beside me, high above the surrounding lowlands, I could see landslides chewing away at the scraps of the ancient plateau. On all sides of this jungle-covered mesa, erosion was stripping off a hundred million years' worth of rotted rock along with the deepest soil I'd ever seen.

Since the time of the dinosaurs, water dripping through the equatorial jungle and leaching into the ground has created a deep zone of weathered rock extending hundreds of feet down to the base of the plateau. After South America split off from Africa, the resulting escarpment swept inland eating into the ancient uplands from the side. Standing on the cliff at the edge of the plateau—a small remnant of the original land surface—I admired the wake of new rolling lowlands that fell away toward the Atlantic Ocean.

The Carajás Plateau is made up of banded iron—almost pure iron ore deposited by an anoxic sea long before Earth's oxygen-rich atmosphere evolved. Buried deep in the earth's crust and eventually pushed back to the surface to weather slowly, the iron-rich rock gradually lost nutrients and impurities to seeping water, leaving behind a deeply weathered iron crust.

Aluminum and iron ore can form naturally through this slow weathering process. Over geologic time, the ample rainfall and hot temperatures of the tropics can concentrate aluminum and iron as chemical weathering leaches away almost everything else from the original rock. Although it may take a hundred million years, it is far more cost-effective to let geologic processes do the work than to industrially concentrate the stuff. Given time, this process can make a commercially viable ore—as long as weathering outpaces erosion. If erosion occurs too rapidly, the weathered material disappears long before it could become concentrated enough to be worth mining.

On top of the Carajás Plateau, a gigantic pit opened a window into the earth, extending hundreds of feet to the base of the deep red weathered rock. Huge, three-story-tall trucks crawled up the terraced walls, dragging tons of dirt along the road that snaked up from the bottom. Viewed from the far side, the hundred-foot-tall trees left standing on the rim of the pit looked like a fringe of mold. Gazing at this bizarre sight in the midday sun, I realized how the thin film of soil and vegetation covering Earth's surface resembles a coating of lichen on a boulder.

Speeding off the plateau, we dropped down to the young rolling hills made of rock that once lay beneath the now-eroded highlands. As we drove through virgin rainforest, road cuts exposed soil one to several feet thick on the dissected slopes leading down to the deforested lowland. Leaving the jungle, we saw bare slopes that provided stark evidence that topsoil erosion following forest clearing led to abandoned farms. Around villages on the forest's edge, squatters farmed freshly cleared tracts. Weathered rock exposed along the road poked out of what had until recently been soil-covered slopes. The story was transparently simple. Soon after forest clearing, the soil eroded away and people moved deeper into the jungle to clear new fields.

A few miles in from the forest edge, family farms and small villages gave way to cattle ranches. As subsistence farmers pushed farther into the forest, ranchers took over abandoned farms. Cows can graze land with soil too poor to grow crops, but it takes a lot of ground to support them. Large-scale cattle grazing prevents the forest from regrowing, causing further erosion and sending frontier communities farther and farther into the jungle in an endless push for fresh land. The vicious cycle is plainly laid out for all to see.

Instead of clearing small patches of forest for short periods, immigrants to the Amazon are clearing large areas all at once, and then accelerating erosion through overgrazing, sucking the life from the land. The modern cycle of forest clearing, peasant farming, and cattle ranching strips off topsoil and nearly destroys the capacity to recover soil fertility. The result is that the land sustains fewer people. When they run out of productive soil, they move on. The modern Amazon experience reads a lot more like the history of North America than we tend to acknowledge. Yet the parallel is as clear as it is fundamental.

Between forty million and one hundred million people lived in the Americas when Columbus "discovered" the New World—some four million to ten million called North America home. Native Americans along the East Coast practiced active landscape management but not sedentary agriculture. Early colonists described a patchwork of small clearings and the natives' habit of moving their fields every few years, much like early Europeans or Amazonians. While there is emerging evidence of substantial local soil erosion from native agriculture, soil degradation and erosion began to transform eastern North America under the new arrivals' more settled style of land use.

Intensive cultivation of corn quickly exhausted New England's nutrient-poor glacial soils. Within decades, colonists began burning the forest to make ash fertilizer for their fields. With more people crowded into less space, New Englanders ran out of fresh farmland faster than their neighbors in the South. Early travelers complained about the stench from fields where farmers used salmon as fertilizer. And in the South, tobacco dominated the slave-based economies of Virginia and Maryland and soil exhaustion dominated the economics of tobacco cultivation. Once individual family farms coalesced into slave-worked tobacco plantations, the region became trapped in an insatiable socioeconomic system that fed on fresh land.

Historian Avery Craven saw colonial soil degradation as part of an inevitable cycle of frontier colonization. "Men may, because of ignorance or habit, ruin their soils, but more often economic or social conditions, entirely outside their control lead or force them to a treatment of their lands that can end only in ruin."[1] Craven thought frontier communities generally exhausted their soil because of the economic imperative to grow the highest value crop. The tobacco economy that ruled colonial Virginia and Maryland was exactly what Craven had in mind.

In 1606 James I granted the Virginia Company a charter to establish an English settlement in North America. Founded by a group of London investors, the company expected their New World franchise to return healthy profits. Under the leadership of Captain John Smith, on May 14, 1607, the first load of colonists landed along the banks of the James River sixty miles up Chesapeake Bay. Hostile natives, disease, and famine killed two-thirds of the original settlers before Smith returned to England in 1609.

Desperately searching for ways to survive, let alone earn a profit, the Jamestown colonists tried making silk, then glass; harvesting timber; growing sassafras; and even making beer. Nothing worked until tobacco provided a profitable export that propped up the colony.

Sir Walter Raleigh is often credited with introducing tobacco to England in 1586. Whether or not that dubious honor is actually his, Spanish explorers brought both leaf and seeds back from the West Indies. Smoking became immensely popular and the English developed quite a taste for Spanish tobacco grown with slave labor in the Caribbean. Sold at a premium to London merchants, tobacco offered just what the Jamestown colonists needed to keep their colony afloat.

Unfortunately, England's new smokers did not like Virginian tobacco. With an eye toward competing in the London market colonist John Rolfe (perhaps better remembered as Pocahontas's husband) experimented with planting Caribbean tobacco. Satisfied that the stuff "smoked pleasant, sweete and strong," Rolfe and his compatriots shipped their first crop to England. It was a hit in London's markets, comparing favorably with premium Spanish tobacco.

Soon everybody was planting tobacco. Twenty thousand pounds were sent to England in 1617. Twice as much set sail in the next shipment. Captain Smith praised Virginia's "lusty soyle" and the colonial economy quickly became dependent upon tobacco exports. On September 30, 1619, colonist John Pory wrote to Sir Dudley Carleton that things were finally turning around. "All our riches for the present doe consiste in Tobacco, wherein one man by his owne labour hath in one yeare raised to himselfe to the value of £200 sterling and another by the meanes of six Servants hath cleared at one crop a thousand pound English."[2] Within a decade, one and a half million pounds of Virginian tobacco reached English markets each year.

America's colonial economy was off and running. Within a century, annual exports to Britain soared a thousandfold to more than twenty million pounds. Tobacco so dominated colonial economics that it served as an alternative currency. The stinking weed saved the faltering colony, but growing it triggered severe soil degradation and erosion that pushed colonists ever inland.

Colonial tobacco was a clean-tilled crop. Farmers heaped up a pile of dirt around each plant with a hoe or a light one-horse plow. This left the soil exposed to rainfall and vulnerable to erosion during summer storms that hit before the plants leafed out. Despite the obvious toll on the land, tobacco had a singular attraction. It fetched more than six times the price of any other crop, and could survive the long (and expensive) journey across the Atlantic. Most other crops rotted on the way or could not sell for enough to pay for the trip.

Colonial economics left little incentive to plant a variety of crops when tobacco yielded by far the greatest returns. So Virginians grew just enough food for their families and devoted their energy to growing tobacco for European markets. New land was constantly being cleared and old land abandoned because a farmer could count on only three or four highly profitable tobacco crops from newly cleared land. Tobacco strips more than ten times the nitrogen and more than thirty times the phosphorous from the soil than do typical food crops. After five years of tobacco cultivation the ground was too depleted in nutrients to grow much of anything. With plenty of fresh land to the west, tobacco farmers just kept on clearing new fields. Stripped bare of vegetation, what soil remained on the abandoned fields washed into gullies during intense summer rains. Virginia became a factory for turning topsoil into tobacco.

King James saw the tobacco business as an attractive way to raise revenue. In 1619 the Virginia Company agreed to pay the Crown one shilling per pound on its shipments to England in exchange for restrictions on Spanish tobacco imports and on tobacco growing in England—a monopoly on the popular new drug. Just two years later, new regulations mandated that all tobacco exported from the colonies be sent to England. In 1677 the royal treasury pocketed £100,000 from import duties on Virginian tobacco and another £50,000 from Maryland tobacco. Virginia returned more to the royal pocketbook than any other colony; more than four times the revenue from the East Indies.

Not surprisingly, colonial governments piled on to use tobacco to raise revenue. Once hooked on the new source of cash, they quickly squashed attempts to stem reliance on tobacco. When Virginians requested a temporary ban on tobacco growing in 1662, they were unsubtly told never to make such a request again. The secretaries for the colony of Maryland tried to ensure that colonists did not "turn their thoughts to anything but the Culture of Tobacco."[3]

The short-lived fertility of the land under tobacco cultivation encouraged rapid expansion of agricultural settlements. Abandoning fields that no longer produced adequate returns, Virginia planters first requested permission to clear new land farther from the coast in 1619. Five years later, planters at Paspaheigh sought permission from the colonial court to move to new land, even though fifteen years before their governor had pronounced their lands excellent for growing grain. Little more than two decades later, tobacco farmers along the Charles River petitioned the governor for access to virgin lands because their fields "had become barren from cultivation." Seventeenth-century Virginians complained about the extreme loss of soil during storms; it was hard to overlook the destructive gullies chewing up the countryside. Moving inland, planters encountered soils even more susceptible to erosion than those along the coast and in the major river valleys. Pushing south as well, by 1653 tobacco farmers were clearing new fields in North Carolina's coastal plain where there was still plenty of fresh land.

As soil fertility declined along the coast, farmers moved inland. The potential for access to the rich lands beyond the mountains motivated Virginians during the French and Indian War. Colonial farmers were enraged at mother England when the peace treaty of 1763 effectively closed the western lands to immediate settlement. Lingering resentment over debilitating tobacco taxes and perceived obstructions to westward expansion helped fuel dissatisfaction with British rule.

Colonial agriculture remained focused on tobacco in the South despite depressed prices stemming from oversupply and the requirement that the whole crop be shipped to England. By the middle of the eighteenth century, government duties accounted for about 80 percent of the sale price of tobacco; the planter's share had dropped to less than 10 percent. Anger over perceived inequities in the regulation, sale, and export of tobacco simmered until the Revolutionary War.

Particularly in the South, the ready availability of new lands meant that farmers neglected crop rotation and the use of manure to replenish soils. Published in 1727, *The Present State of Virginia* blamed the rapid decline in soil fertility on failure to manure the fields. "So it is at present that Tobacco swallows up all other Things, every thing else is neglected. . . . By that time the Stumps are rotten, the Ground is worn out; and having fresh Land enough . . . they take but little Care to recruit the old Fields with Dung."[4] Moving on to fresh ground was easier than collecting and spreading manure—as long as there was plenty of land for the taking.

Other contemporary observers also noted that tobacco consumed the full attention of planters. In a 1729 letter to Charles Lord Baltimore, Benedict Leonard Calvert concisely summarized the influence of tobacco on colonial agriculture. "In Virginia and Maryland Tobacco is our Staple, is our All and Indeed leaves no room for anything Else."[5] Tobacco reigned as undisputed king of the southern colonies.

The need for continual access to fresh land encouraged the establishment of large estates. Low prices in the glutted tobacco market of the late seventeenth century created opportunities to consolidate large holdings when small farmers went out of business. Just as in Rome two thousand years before, and in the Amazon almost three centuries later, abandoned family farms ended up in the hands of plantation owners.

In New England, some colonists began experimenting with soil improvement. Connecticut minister, doctor, and farmer Jared Eliot published the first of his *Essays Upon Field Husbandry* in 1748 reporting the results of experiments on how to prevent or reverse soil degradation. Riding on horseback to call on his parishioners and patients, Eliot noticed that the muddy water running off bare hillsides carried away fertile soil. He saw how deposition of mud washed from the hills enriched valley bottom soils and how loss of the topsoil ruined upland fields. Eliot recommended spreading manure and growing clover for improving poor soils. He endorsed marl (fossil sea shells) and saltpeter (potassium nitrate) as excellent fertilizers almost equal to good dung. Bare soil left exposed on sloping ground was particularly vulnerable to washing away in the rain. Sound as it was, few colonial farmers heeded Eliot's advice, particularly in the South where new land was still readily obtained.

Benjamin Franklin was among those who bought Eliot's essays and began experimenting with how to improve his land. Writing to Eliot in 1749 he confided his concern about how difficult it would be to convince American farmers to embrace soil husbandry. "Sir: I perused your two

Essays on Field Husbandry, and think the public may be much benefited by them; but, if the farmers in your neighborhood are as unwilling to leave the beaten road of their ancestors as they are near me, it will be difficult to persuade them to attempt any improvement."[6] Eliot likened farmers who did not return manure and crop wastes to the fields to a man who withdrew money from the bank without ever making a deposit. I imagine that Franklin concurred.

Commentary on the depleted state of colonial soils was routine by the end of the eighteenth century. Writing during the Revolutionary War, Alexander Hewatt described farmers in the Carolinas as focused on short-term yields and paying little attention to the condition of their land.

> Like farmers often moving from place to place, the principal study with the planters is the art of making the largest profit for the present time, and if this end is obtained, it gives them little concern how much the land may be exhausted. . . . The richness of the soil, and the vast quantity of lands, have deceived many. . . . This will not be the case much longer, for lands will become scarce, and time and experience, by unfolding the nature of the soil . . . will teach them . . . to alter [their] careless manner of cultivation.[7]

Hewatt was not alone in his bleak assessment of American agriculture.

Many Europeans who traveled through the southern states in the late 1700s expressed surprise at the general failure to use manure as a soil amendment. Exiled French revolutionary Jacques-Pierre Brissot de Warville toured the newly independent United States in 1788 and wondered at the ruinous style of agriculture. "Though tobacco exhausts the land to a prodigious degree, the proprietors take no pains to restore its vigour; they take what the soil will give, and abandon it when it gives no longer. They like better to clear new lands, than to regenerate old. Yet these abandoned lands would still be fertile, if they were properly manured and cultivated."[8] Careless waste of good land perplexed European observers accustomed to cheap labor and a shortage of fertile land.

At the close of the eighteenth century, newly arrived settler John Craven found Virginia's Albemarle County so degraded by poor farming practices that the inhabitants faced the simple choice of emigrating or improving the soil. Writing to the *Farmer's Register* years later, Craven recalled the sad state of the land. "At that time the whole face of the country presented a scene of desolation that baffles description—farm after farm had been worn out, and washed and gullied, so that scarcely an acre could be found in a place fit for cultivation. . . . The whole of the virgin soil was washed and carried off from the ridges into the valleys."[9] Visiting Virginia and Maryland the following year, in 1800, a baffled William Strickland declared that he could not see how the inhabitants scratched a living from their fields.

In 1793 Unitarian minister Harry Toulmin left Lancashire for America to report to his congregation on the suitability of the new country for emigration. Land hunger and the rising price of food in Britain increased pressure to leave for America, particularly for those living on fixed incomes and low wages in the industrializing economy. In addition, many Unitarians and others sympathetic to the progressive ideals of the American and French revolutions abandoned their homeland for the New World when the new French Republic declared war on England.

Finding agricultural prospects poor along the Atlantic seaboard, Toulmin procured letters of introduction from James Madison and Thomas Jefferson to John Breckinridge, who had quit a Virginia congressional seat to emigrate to Kentucky. Toulmin's letters and journals provide a vivid description of Kentucky's soils at the time of first settlement. Reporting on the agricultural potential of Mason County in northern Kentucky, Toulmin described the gently undulating country as well endowed with rich soil. "The soil is in general rich loam. In the first-rate land (of which there are some million of acres in this county) it is black. The richest and blackest mold continues to about the depth of five or six inches. Then succeeds a lighter colored, friable mold which extends about fifteen inches farther. When dry it will blow away with the wind."[10] Testimony such as Toulmin's helped draw people west from the coast. It also proved far more prophetic than he could have imagined.

About the time of the American Revolution, some of the founding fathers began to worry about the impact of mining the soil on the country's future. George Washington and Thomas Jefferson were among the first to warn of the destructive nature of colonial agriculture. Ideological rivals, these prosperous Virginia plantation owners shared concern over the long-term effects of American farming practices.

After the Revolution, Washington did not hide his scorn for the shortsighted practices of his neighbors. "The system of agriculture, (if the epithet of system can be applied to it) which is in use in this part of the United States, is as unproductive to the practitioners as it is ruinous to the landholders."[11] Washington blamed the widespread practice of growing tobacco for wearing out the land. He saw how poor agricultural practices fueled the desire to wrest the greatest return from the ground in the shortest time—and vice versa. In a 1796 letter to Alexander Hamilton, Washington predicted that soil exhaustion would push the young country inland. "It must be obvious to every man, who considers the agriculture of this country . . . how miserably defective we are in the management of [our lands]. . . . A few years more of increased sterility will drive the Inhabitants of the Atlantic States westward for support; whereas if they were taught how to improve the old, instead of going in pursuit of new and productive soils, they would make these acres which now scarcely yield them any thing, turn out beneficial to themselves."[12]

Washington's interest in progressive agriculture began long before the Revolution. As early as 1760, he used marl (crushed limestone), manure, and gypsum as fertilizers and plowed crops of grass, peas, and buckwheat back into his fields. He built barns for cattle in order to harvest manure, and instructed reluctant plantation managers to spread the waste from livestock pens onto the fields. He experimented with crop rotations before finally settling on a system that involved interspersing grains with potatoes and clover or other grasses. Washington also experimented with deep plowing to reduce runoff and retard erosion. He filled gullies with old fence posts, trash, and straw before covering them with dirt and manure and then planting them with crops.

Perhaps most radical, however, was Washington's realization that soil improvement was next to impossible on large estates. Dividing his land into smaller tracts, he instructed his overseers and tenants to promote soil improvement. Washington's efforts focused on preventing soil erosion, saving and using manure as fertilizer, and specifying cover crops to include in rotations.

Returning to Mount Vernon after the Revolution, Washington wrote English agriculturalist Arthur Young for advice on improving his lands. Young embraced Washington as a "brother farmer" and agreed to provide the American president with any assistance desired.

In 1791 Young asked Washington to describe agricultural conditions in northern Virginia and Maryland. Washington's reply indicates that the old practices that encouraged soil erosion and exhaustion remained widespread. In particular, the practice of growing steadily falling yields of tobacco, followed by as much corn as the exhausted land could produce, continued to reduce soil fertility. With limited pasture and livestock, few farmers used manure to prolong or restore soil fertility. Washington explained that American farmers had a strong incentive to get the most out of their laborers regardless of the effect on their soil; workers cost four times the value of the land they could work. He also reported a growing tendency to abandon tobacco in favor of wheat, even though wheat yields were barely comparable to medieval European yields. American agriculture was wearing down the New World.

Thomas Jefferson too worried that Americans were squandering the productive capacity of their land. Where Washington blamed ignorance of proper farming methods, Jefferson saw greed. "The indifferent state of [agriculture] among us does not proceed from a want of knowledge merely; it is from our having such quantities of land to waste as we please. In Europe the object is to make the most of their land, labor being abundant; here it is to make the most of our labor, land being abundant."[13] When a perplexed Arthur Young questioned how a man could produce five thousand bushels of wheat on a farm with cattle worth only £150, Jefferson reminded him that "manure does not enter into this because we can buy an acre of new land cheaper than we can manure an old one."[14] Better short-term returns were to be had by mining soil than by adopting European-style husbandry. In Jefferson's view, failure to care for the land was the curse of American agriculture.

The relationship between eighteenth-century plantation owners and their poorer neighbors bolsters Jefferson's argument. Wealthy landowners generally exhausted their land growing tobacco, used their slaves to clear new fields, and then sold their old fields to farmers lacking the means and the slaves to clear and work a tobacco plantation. Plantation owners often bought food from neighboring farms to feed their own families. Cotton and tobacco so dominated agriculture that before the Civil War the South was a net importer of grains, vegetables, and farm animals.

In the spring of 1793 Jefferson's son-in-law Colonel T. M. Randolph started plowing horizontally along hillslope contours rather than straight downhill. Skeptical at first, Jefferson himself became a convert when Randolph developed a hillside plow fifteen years later. Thereafter a vocal proponent of contour plowing, Jefferson testified that formerly erosive rains no longer carved deep gullies across his fields. Randolph's invention won him a prize from the Albemarle County Agricultural Society in 1822. Together with his famous father-in-law, Randolph popularized contour plowing through an extensive network of correspondents.

In one such letter, written to C. W. Peale in 1813, Jefferson extolled the virtues of the new practice.

Our country is hilly and we have been in the habit of ploughing in straight rows whether up and down hill, in oblique lines, or however they lead; and our soil was all rapidly running into the rivers. We now plough horizontally, following the curvatures of the hills and hollows, on the dead level, however crooked the lines may be. Every furrow then acts as a reservoir to receive and retain the waters, all of which go to the benefit of the growing plant, instead of running off into the streams. In a farm horizontally and deeply ploughed, scarcely an ounce of soil is carried off from it.[15]

But the new approach had to be employed with care. Even if pitched slightly down slope, furrows would still collect runoff and guide incipient streams into gullies. Though contour plowing caught on, many considered the effort needed to do it, let alone do it right, too much bother. In the 1830s Randolph's son described how the "new" practices of deep plowing, fertilizing with gypsum, and rotating corn with clover or grass would soon eclipse his father's contribution in the fight to reclaim worn out lands.

Early in the nineteenth century, Americans began to recognize the need to safeguard and restore soil fertility. Some farmers began plowing deeper and adding animal and vegetable manures to their fields. In particular, agriculturalist John Taylor argued that soil conservation and improvement were necessary to sustain southern agriculture. "Apparent to the most superficial observer, is, that our land has diminished in fertility. . . . I have known many farms for above forty years, and . . . all of them have been greatly impoverished." Forecasting the future of the South, Taylor predicted "our agricultural progress, to be a progress of emigration,"[16] unless soil improvement became the region's agricultural philosophy. By the 1820s, the need for aggressive efforts to improve the soil was widely recognized throughout the South.

Taylor's French contemporary Félix de Beaujour characterized American farmers as nomads continually on the move. He marveled at their general reluctance to use manure to restore soil fertility. "The Americans appear to be ignorant that with water manure is every where made; and that with manure and water, there is not an inch of ground that cannot be made fertile. The land for this reason is there soon exhausted, and . . . the farmers of the United States resemble a people of shepherds, from their great inclination to wander from one place to the other."[17] Such descriptions abound in early nineteenth-century accounts of the South.

Rural newspapers across the country carried the remarks of retired president James Madison on the front page when he addressed Virginia's Albemarle County Agricultural Society in May 1818. Madison cautioned that the nation's westward expansion did not necessarily mean progress. Building a nation with a future required caring for and improving the land. Neglect of manure, working the land too hard by cultivating it continuously, and plowing straight up and down hills would rob land of its fertility. Madison cautioned that agricultural expansion be moderated; improving the soil was not just an alternative to heading farther west, it was the only option over the long run.

The ideas of Pennsylvania farmer John Lorain, whose book *Nature and Reason Harmonized in the Practice of Husbandry* was published after his death in 1825, maintained that erosion was beneficial under natural vegetation because soil gained as much as it lost. Valley bottoms were enriched by soil eroded from hillsides where weathering produced new soil that replaced the eroded material. Farmers changed the system so that erosion from improvident use of the plow and exposure of bare soil to rain impoverished both the soil and the people working the land.

Lorain suggested using grass as a permanent crop on steeply sloping land and putting fields to pasture before they became exhausted. The grass cover would prevent erosion by intercepting and absorbing the impact of rainfall and keep the ground porous enough for precipitation to sink into the ground, instead of running off over the surface. The key to preventing erosion and maintaining soil fertility was to incorporate as much vegetable and animal matter into the soil as possible. An advocate of inexpensive erosion control measures that even poor farmers could adopt, he maintained that careful attention to plowing along contours and preventing surface runoff from gathering into destructive gullies could conserve the soil.

Lorain also saw the tenant system as a major obstacle to soil conservation. The novel efforts of gentleman farmers like Washington and Jefferson discouraged small farmers who could not afford the expense. Instead, tenants with no vested interest in the land wasted the soil, and ignored potentially beneficial conservation measures. His solution was to free his slaves and mandate soil improvement as a condition of all leases. Lorain ridiculed the idea held by many farmers that they could find, somewhere, a soil that was inexhaustible. "When the Pacific Ocean puts a stop to their progress, it is possible they will be convinced, that no such soil exists."[18]

Many other contemporary observers who examined the question of soil exhaustion concluded that lack of manure was to blame for rapid exhaustion of the region's soil. Using slaves to grow livestock fodder was far less profitable than applying their labor toward cultivating cotton or tobacco.

Although it was known that a well-manured field would produce two to three times the harvest of unmanured fields, Southerners left their cattle to graze in the woods year round. On most plantations, virtually no effort was made to gather dung and spread it on the fields; numerous historical accounts refer to the sad state of southern cattle.

An article in the October 11, 1827, edition of the *Georgia Courier* quoted a letter from a traveler passing through Georgia who noted that the exhausted land provided common motivation for a steady stream of westbound emigrants. "I now left Augusta; and overtook hordes of cotton planters from North Carolina, South Carolina, and Georgia, with large groups of negroes, bound for Alabama, Mississippi, and Louisiana; 'where the cotton land is not worn out.'"[19] The South was heading West.

By the 1820s slavery was becoming less economically viable along the eastern seaboard. John Taylor noted that many plantation owners refused to abandon even marginally profitable tobacco cultivation because to do so would have left them without winter work for their slaves. As much land lay abandoned in North Carolina as was being farmed. Low prices for tobacco and cotton, owing to competition from farms working fresh soils to the west, kept profits low on the depleted Piedmont and coastal lands. Slaves began to be a burden to their masters. On March 24, 1827, the *Niles Register* complained about the situation. "Most of our intelligent planters regard the cultivation of tobacco in Maryland as no longer profitable and would almost universally abandon it if they knew what to do with their slaves."[20]

Emigrant planters continued their destructive ways in the new western lands to which their old habits had driven them. Writing to *Farmer's Register* in August 1833, one Alabama resident expressed dismay at continuing the cycle. "I have not much hope of seeing improvement in the agriculture of this state. Our planters are guilty of the same profligate system of destroying lands that has characterized their progenitors of Georgia, the Carolinas, and Virginia, immemorially. They wage unmitigated war both against the forest and the soil—carrying destruction before them, and leaving poverty behind."[21] There was no debate about the connection between abused land and depressed economies in nineteenth-century America. A nation of farmers could read the signs for themselves.

As the editor of the *Cultivator*, Jesse Buel was the most articulate representative of conservative farmers who embraced agricultural improvement rather than westward emigration. Born in Connecticut two years after the opening salvos of the Revolution, Buel apprenticed to a printer and then purchased a farm in the 1820s. A decade later he began championing manure as the key to rural prosperity, believing that land judiciously managed need not wear out. In his view, it was a farmer's duty to treat the land as a trust to be passed on unimpaired to posterity.

Buel's views were shared by German and Dutch farmers who immigrated to Pennsylvania, bringing progressive European agricultural practices that contrasted with typical colonial practices. They organized their modest farms around giant barns where cows turned fodder crops into milk and manure. Unlike most American farmers, they treated dirt like gold. Their land prospered, yielding bountiful harvests that astounded visitors from the South where publication of Edmund Ruffin's *Essay on Calcareous Manures* in 1832 initiated a revolution in American agriculture.

Better known to history as an early agitator for southern independence, Ruffin believed in the power of agrochemistry to restore soil fertility—and the South. Ruffin inherited a rundown family plantation in 1810 at the age of sixteen. Struggling to turn a profit from fields already farmed for a century and a half, he adopted the deep plowing, crop rotation, and grazing exclusion advocated by agricultural reformer John Taylor. Unimpressed with the results and almost ready to join the exodus westward, Ruffin tried applying marl to his land.

The results were dramatic. Plowing crushed fossil shells into his fields raised corn yields by almost half. Ruffin began adding marl to more of his land and almost doubled his wheat crop. Concluding that Virginia's soils were too acidic to sustain cultivation, Ruffin reasoned that adding calcium carbonate to neutralize the acid would enable manure to sustain soil fertility. His essay received widespread attention and favorable reviews in leading agricultural journals.

Following Ruffin's example, Virginia's farmers began increasing their harvests. Propelled to prominence in southern society, Ruffin began publishing *Farmer's Register,* a monthly journal devoted to the improvement of agriculture. The newspaper carried no advertisements and featured practical articles written by farmers. Within a few years Ruffin had more than a thousand subscribers. Eager to compete with the new cotton kingdom emerging out West, South Carolina's newly elected governor James Hammond hired Ruffin to locate and map the state's marl beds in 1842. Ten years later Ruffin accepted the presidency of the newly formed Virginia Agricultural Society.

Well known, highly regarded, and with a lust for public attention, Ruffin turned his attention to advocating southern independence in the 1850s.

Seeing secession as the only option, he argued that slave labor had sustained advanced civilizations like ancient Greece and Rome. Upon learning of Lincoln's election, Ruffin hastened to attend the convention that adopted the ordinance of secession. When the sexagenarian was awarded the distinction of firing the first shot at Fort Sumter in April 1861, he had already helped start an agrochemical revolution by demonstrating that manipulating soil chemistry could enhance agricultural productivity.

Ruffin thought that soils were composed of three major types of earths. Siliceous earths were the rock minerals that allowed water to pass freely and were thus the key to a well-drained soil. Aluminous earths (clays) absorbed and retained water, creating networks of cracks and fissures that served as miniature reservoirs. Calcareous earths could neutralize acidic soils. Ruffin thought that soil fertility lay in the upper few inches of a soil where organic material mixed with the three earths. Productive agricultural soils were those composed of the right combination of siliceous, aluminous, and calcareous earths.

Ruffin recognized that topsoil erosion squandered soil fertility. "The washing away of three or four inches in depth, exposes a sterile subsoil . . . which continues thenceforth bare of all vegetation."[22] He also agreed with agricultural authorities that manure could help revive the South. But he thought that the ability of manure to enrich soil depended on a soil's natural fertility. Manure would not improve harvests from acidic soils without first neutralizing the acid. Ruffin did not believe that calcareous earth fertilized plants directly; supplemented by calcareous earth, manure could unleash masked fertility and transform barren ground back into fertile fields.

Ruffin further saw that the institution of slavery made the South dependent on expanding the market for slaves born on plantations. He believed that surplus slaves had to be exported unless agricultural productivity could be increased enough to feed a growing population. Ruffin's views on agricultural reform and politics collided with the reality of the Civil War. He committed suicide shortly after Lee surrendered.

The problem of soil exhaustion was not restricted to the South. By the 1840s, addresses to agricultural societies in Kentucky and Tennessee warned that the new states were rapidly emulating Maryland and Virginia in squandering their productive soils. By the advent of mechanized agriculture in the mid-nineteenth century, per-acre wheat yields in New York were just half of those from colonial days despite advances in farming

Figure 14. Charles Lyell's illustration of a gully near Milledgeville, Georgia, in the 1840s (Lyell 1849, fig. 7).

methods. Still, the more diversified northern economy made the effects of soil depletion on northern states less pronounced than in the South.

In the 1840s British geologist Charles Lyell toured the antebellum South, stopping to investigate deep gullies gouged into the recently cleared fields of Alabama and Georgia. Primarily interested in the gullies as a way to peer down into the deeply weathered rocks beneath the soil, Lyell noted the rapidity with which the overlying soil eroded after forest clearing. Across the region, the consistent lack of evidence for prior episodes of gully formation implied a fundamental change in the landscape. "I infer, from the rapidity of the denudation caused here by running water after the clearing or removal of wood, that this country has been always covered with a dense forest, from the remote time when it first emerged from the sea."[23] Lyell saw that clearing the rolling hills for agriculture had altered an age-old balance. The land was literally falling apart.

One gully in particular attracted Lyell's attention. Three and a half miles west of Milledgeville on the road to Macon, it began forming in the 1820s, when forest clearing exposed the ground to direct assault by the elements. Monstrous three-foot-deep cracks opened up in the clay-rich soil during the summer. The cracks gathered rainwater and concentrated erosive runoff, incising a deep canyon. By Lyell's visit in 1846, the gully had grown into a chasm more than fifty feet deep, almost two hundred feet wide, and three hundred yards long. Similar gullies up to eighty feet deep had consumed recently cleared fields in Alabama. Lyell considered the rash of gullies a serious threat to southern agriculture. The soil was washing away much faster than it could possibly be produced.

Passing through an area of low rolling hills on the road to Montgomery, Lyell marveled over the stumps of huge fir trees in a recent clearing. Curious as to how many years it would take to regrow such a forest, he measured the diameter of stumps and counted their annual growth rings. The smallest spanned almost two and a half feet in diameter with a hundred and twenty annual rings; the largest was four feet in diameter and had three hundred and twenty rings. Lyell was confident that such venerable trees could never regrow in the altered landscape. "From the time taken to acquire the above dimensions, we may confidently infer that no such trees will be seen by posterity, after the clearing of the country, except where they may happen to be protected for ornamental purposes."[24] Tobacco, cotton, and corn were replacing the forest of immense trees that had covered the landscape for ages. Bare and exposed, the virgin soil bled off the landscape with every new storm.

In addition to impressive gullies, Lyell ran into families abandoning their farms and moving to Texas or Arkansas. Passing thousands of people migrating westward, Lyell reported that those he met kept asking, "Are you moving?" After showing the eminent geologist some fossils, one elderly gentleman offered to sell Lyell his entire estate. Lyell pressed him as to why he was so eager to sell the land he had cleared himself and lived on for twenty years. The man replied: "I hope to feel more at home in Texas, for all my old neighbours have gone there."[25]

Traveling through much of the South by canoe, Lyell watched the rivers along the way, describing how dramatically accelerated soil erosion following forest clearing and cultivation were obvious to anyone paying attention. Special training in geology was not needed to read the signs of catastrophic erosion. People he met along Georgia's Alatamaha River told him that the river had flowed clean even during floods until the land upriver was cleared for planting. As late as 1841, local residents could determine the source of floodwaters from individual storms because the deforested branch of the river ran red with mud, while the still forested branch flowed crystal clear even during big storms. By the time of Lyell's visit the formerly clear branch also flowed muddy after Native Americans were driven out and the land was cleared for agriculture.

The toll of contemporary agricultural methods on soil and society was no secret. The report of the commissioner of patents for 1849 attempted to tally up the cost to the country.

> One thousand millions of dollars, judiciously expended, will hardly restore the one hundred million acres of partially exhausted lands in the Union to that richness of mould, and strength of fertility for permanent cropping, which they possessed in their primitive state. . . . Lands that, seventy years ago, produced from twenty-five to thirty-five bushels of wheat in the State of New York, now yield only from six to nine bushels per acre; and in all the old planting States, the results of exhaustion are still more extensive and still more disastrous.[26]

Since falling crop yields were apparent throughout the original states, how to protect soil fertility presented a fundamental challenge. "There appears to be no government that realizes its duty 'to promote the public welfare' by . . . impressing upon them the obligation which every cultivator of the soil owes to posterity, not to leave the earth in a less fruitful condition than he found it."[27] Before the start of the Civil War, agricultural periodicals throughout the country assailed the twin evils of soil erosion and exhaustion. As the shortage of fresh land became acute, pleas to adopt soil conservation and improvement techniques became increasingly common.

The immediate causes of soil exhaustion in the antebellum South were not mysterious. Foremost among these were continuous planting without crop rotation, inadequate provision for livestock to provide manure, and improvident tilling straight up and down sloping hillsides that left bare soil exposed to rainfall. But there were underlying social causes that drove these destructive practices.

There can be no doubt that the desire for the greatest short-term returns drove plantation agriculture. Land was cheap and abundant. Moving farther inland every few years, a planter could enjoy the benefits of perpetually farming virgin ground—as long as there was new ground to be had. Clearing new fields was cheap compared to carefully plowing, terracing, and manuring used land. Still, finding virgin land required uprooting and relocating the family and all its possessions, including slaves, to newly opened states in the West. Given the high cost of moving—both socially and financially—what kept such practices alive in the face of overwhelming evidence they were ruining the land?

For one thing, the large plantations' owners—those most likely to recognize the problem of soil exhaustion—did not work their own land. Just as two thousand years before in ancient Rome, absentee ownership encouraged soil-wasting practices. Overseers and tenant farmers paid with a percentage of the crop were more concerned about maximizing each year's harvest than protecting the landowner's investment by maintaining soil fertility. Time invested in plowing along contours, repairing nascent gullies, or delivering manure to the fields reduced their immediate income. Overseers who rarely remained on the same ground for more than a year skimmed off a farm's fertility as quickly as possible.

Another fundamental obstacle to agricultural reform was that the institution of slavery was incompatible with methods for reversing soil degradation. In a way, the intensity of soil erosion in the antebellum South helped trigger the Civil War. While we're all taught that the Civil War was fought over slavery, what we don't learn is that the tobacco and cotton monocultures that characterized the southern economy required slave labor to turn a profit. More than a cultural convention, slavery was essential to the underpinnings of southern wealth. It was not simply that the South was agricultural; much of the North was too. Slavery was critical to the export-oriented, cash-crop monoculture common throughout the South.

Of course, any comprehensive explanation for the Civil War must address a complex set of conditions and events that predated the outbreak of hostilities. The main reasons for the Civil War are usually given as controversy over tariffs and the establishment of a central bank, abolitionist agitation both in Congress and the North in general, and passage of fugitive slave laws. Obviously, efforts to outlaw slavery arose from its ongoing practice in the South. But the most volatile issue of the period preceding the Civil War was the question of slavery's status in the new western states.

Tensions came to a head after the Supreme Court's infamous 1857 Dred Scott decision that slaves were not citizens and therefore lacked standing to sue for their freedom. Five of the nine Supreme Court justices came from slave-holding families. Pro-slavery presidents from southern states had appointed seven. The decision declared the 1820 Missouri Compromise unconstitutional, holding that the federal government had no authority to restrict slavery in the new territories. Southerners hailed an apparent vindication of their views.

Outraged northern abolitionists embraced the upstart Republican Party and after much politicking nominated long-shot candidate Abraham Lincoln for president on a platform that held slavery should spread no farther. The Democrats fragmented when Northerners endorsed Stephen Douglas and Southerners broke ranks to nominate Vice President John Breckinridge of Kentucky. The Constitutional Union Party composed of diehard Whigs from the border states nominated John Bell of Tennessee.

Fragmented opposition was just what Lincoln needed. In an election split along geographical lines, the southern states went for Breckinridge. The border states of Kentucky, Virginia, and Tennessee voted for Bell. Douglas carried Missouri and New Jersey. Lincoln won just 40 percent of the popular vote, but carried a majority of electoral votes—all the northern states plus the new states of California and Oregon.

With Lincoln in the White House, war became increasingly likely. Northern perspectives leading up to the war are easy to grasp. Abolitionists considered slavery immoral. Many Northerners regarded legalized slavery as inconceivable in a nation based on the precept that all men are created equal. Still, even though Northerners overwhelmingly desired immediate abolition, most were pragmatically content to prevent slavery's expansion into the new territories.

Southern perspectives as war loomed were more complicated, equally pragmatic, and less flexible. Most Southerners believed that Lincoln's election spelled the end of slavery—or at least the end of its expansion to the West. Many were angry over northern interference in their way of life and intrusion into what they considered affairs of their property. Some were incensed about perceived insults to southern honor. But given that Lincoln's election only meant limitation—outright abolition stayed off the table until the war—and that less than a quarter of Southerners actually owned slaves, why did this issue generate enough political friction to blow the country apart?

As is often the case, insight comes if we follow the money. The economic significance of limiting slavery's expansion lies in the central role of soil exhaustion in shaping plantation agriculture and the southern economy.

Most parents of teenagers know that involuntary labor rarely produces quality results. It is hardly surprising that even the best slaves generally do not exhibit initiative, care, and skill. Instead, slaves generally want to maintain competence sufficient to avoid corporal punishment. They cannot be fired from their job and have no incentive to do it well. The very nature of servitude discourages creative expression or expertise at work.

Agriculture tailored to fit the needs of the land requires close attention to detail and flexibility in running a farm. Absentee landlords, hired overseers, and forced labor do not. Furthermore, an adversarial labor system maintained by force necessarily concentrates workers in one place. Single-crop plantation farming thus lent itself well to the rules and routine procedures of slave labor. At the same time, slaves were most profitable when following a simple routine year after year.

Until the 1790s plantations worked by slave labor grew virtually nothing but tobacco. Slave labor became less economical as southern plantations began raising a greater diversity of crops and kept more livestock at the end of the eighteenth century. Many in the South thought that slavery would fade into economic oblivion until the rise of cotton breathed new life into the slave trade. Cotton was almost as hard on the land and relied even more on slave labor than did tobacco.

Slave labor virtually required single-crop farming that left the ground bare and vulnerable to erosion for much of the year. Reliance on a single crop precluded both crop rotation and developing a stable source of manure. If nothing but tobacco or cotton were grown, livestock could not be supported because of the need for grain and grass to feed the animals. Once established, slavery made monoculture an economic necessity—and vice versa. In the half century leading up to the Civil War, southern agriculture's reliance on slave labor precluded the widespread adoption of soil-conserving methods, virtually guaranteeing soil exhaustion.

In contrast to the South, New England's agriculture was more diversified from the start because no lucrative export crop grew there. The fact that slavery did not persist into the late eighteenth century in the northern states may have less to do with abstract ideals of universal freedom and human dignity than the simple reality that tobacco could not grow that far north. Without the continued dominance of large-scale monoculture, slavery might have died out in the South not long after it did in the North.

But this still doesn't explain the vitriolic southern opposition to Lincoln's proposed territorial limitation on the spread of slavery. After all, slavery in the South was not itself directly at issue in the election of 1860. Consider that slaves moved west along with their owners. At the time of the first national census in 1790, Maryland, Virginia, and the Carolinas held 92 percent of all the slaves in the South. Two decades later, after a ban on importing more slaves, the coastal states still held 75 percent of the South's slaves. By the 1830s and 1840s many of the slaveholders in the Atlantic states were breeding slaves for western markets. For plantation owners who stayed behind to work exhausted fields, the slave trade became an economic lifeboat. In 1836, more than one hundred thousand slaves were shipped out of Virginia. One contemporary source estimated that in the late 1850s slave breeding was the largest source of prosperity in Georgia. Census data for 1860 suggest that the value of slaves directly accounted for almost half the value of all personal property in the South, including land. By the start of the Civil War, almost 70 percent of the slaves in the South toiled west of Georgia.

Whether Missouri, Texas, and California would become slave states was a make-or-break issue for plantation owners moving west. The labor-intensive plantation economy of the South required conscripted labor. And for all practical purposes, the rapid soil erosion and soil exhaustion produced by slave-based agriculture condemned the institution of slavery to continuous expansion or collapse. So if slavery was banned in the West, slaves would lose their value—wiping out half of the South's wealth. Lincoln's election threatened slave owners with financial ruin.

Plantation owners knew that new states could create new markets for slaves and their offspring. It was widely expected that allowing slaveholding in Texas would double the value of slaves. The territorial expansion of slavery was a trigger issue for the Civil War because of its immense economic importance to the landed class of the South. While moral issues were hotly debated, friction between the states ignited only after election of a president committed to limiting slavery's expansion.

Whether or not you believe this argument, you don't need to take it on faith that colonial agriculture caused extensive erosion on the eastern seaboard. You can read the evidence for it in the dirt. Soil profiles and valley bottom sediments allow reconstructing the intensity, timing, and extent of colonial soil erosion in eastern North America. Instead of the thick, black topsoil described by the earliest arrivals from Europe, the modern A horizons are thin and clayey. In some places the topsoil is miss-

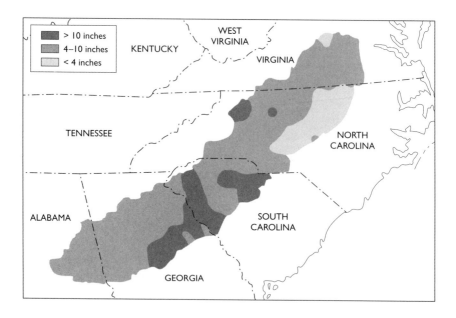

Figure 15. Map of the Piedmont region of the southeastern United States showing the net depth of topsoil eroded from colonial times to 1980 (modified from Meade 1982, 241, fig. 4).

ing entirely, exposing subsoil at the ground surface. Some formerly cultivated parts of the Piedmont have even lost all of their soil, leaving weathered rock exposed at the surface. Soil erosion accelerated by at least a factor of ten under European land use in the colonial era.

Evidence for colonial-era soil erosion is apparent all along the eastern seaboard. Estimates of the average depth of soil erosion in the Piedmont range from three inches to more than a foot since colonial forest clearing. Truncated upland soils missing the top of their A horizon indicate four to eight inches of topsoil loss since colonial farmers began migrating inland. Soils of the southern Piedmont from Virginia to Alabama lost an average of seven inches. Upland soils across two-thirds of the Georgia Piedmont lost between three and eight inches. A century and a half of agriculture in the Carolina Piedmont stripped off six inches to a foot of topsoil. Accelerated erosion was particularly bad under colonial land use, and the problem remains significant today. Sediment yields from forested and agricultural lands in the eastern United States show that agricultural lands still lose soil four times faster than forested land.

Figure 16. North Carolina gully system circa 1911 (Glenn 1911, pl. iiib).

The social and economic impacts of colonial soil erosion were not limited to the farmers who kept moving to find new land to grow tobacco. Just as in ancient Greece and Rome, coastal ports became choked with sediment. Most colonial port towns were located as far inland as possible to minimize overland transport of tobacco. These locations, however, bore the brunt of accelerated soil erosion when material stripped from hillsides reached the estuaries. Half a century of upstream farming converted many open-water ports to mud flats. John Taylor noted that silt washed from hillsides by upland farming buried the bottomlands, filling in coastal rivers and streams and plugging estuaries. At a time when rivers were the nation's highways, the sediments washed from hillsides into rivers and ports were everybody's problem.

Maryland's colonial ports of Joppa Town and Elk Ridge, located on opposite sides of Baltimore, were abandoned after they could no longer accommodate ocean-going vessels. Established by an act of the Maryland legislature in 1707, Joppa Town rapidly grew to become the most important seaport in the colony. The largest oceangoing merchant ships loaded at its wharf until clearing of the uplands started a cycle of erosion that began filling in the bay. By 1768, the county seat was moved to Baltimore where the harbor was unaffected by sedimentation. In the 1940s the remnants of the old wharf stood behind a hundred feet of tree-covered land that extended out past where tall ships once anchored.

The head of Chesapeake Bay shoaled by at least two and a half feet between 1846 and 1938 with deposition of dirt from the surrounding farmlands. The bay also filled with sediment at the head of navigation on the Potomac. A decade after Georgetown was established in 1751, the town built a public wharf extending sixty feet out into deep water. In 1755 a British fleet of heavy warships moored in the river above Georgetown but by 1804 sediment had filled in the main shipping channel.

Arguments over the cause complicated decades of congressional debate over what to do about silting up of the Potomac. By 1837 the river was less than three feet deep above Long Bridge; some blamed construction of the bridge while others blamed the Georgetown causeway. Leading a survey of bridges across the Potomac in 1857, engineer Alfred Rives recognized the true cause as rapid erosion from the extensively plowed hillsides of the surrounding country. Today the Lincoln Memorial sits on ground where ships sailed in the eighteenth century, and Indian missionary Father Andrew White's 1634 description of a crystal-clear Potomac River reads like fiction. "This is the sweetest and greatest river I have seene, so that the Thames is but a little finger to it. There are noe marshes or swampes about it, but solid firme ground. . . . The soyle . . . is excellent . . . commonly a blacke mould above, and a foot within ground of a readish colour. . . . It abounds with delicate springs which are our best drinke."[28]

Sedimentation rates in the Furnace Bay tributary of Chesapeake Bay, just east of the mouth of the Susquehanna River increased almost twentyfold after European settlement. At Otter Point Creek, Maryland, the sedimentation rate in a tidal freshwater delta in upper Chesapeake Bay increased by a factor of six after 1730, and then again by another factor of six by the mid-1800s. The rate of sediment accumulation in a bog at Flat Laurel Gap in the Blue Ridge Mountains of North Carolina remained relatively steady for more than three thousand years and then increased four- to fivefold when land clearance reached the crest of the range around 1880.

Slavery wasn't the only reason that erosion was a bigger problem in the South than in the North. Bare fields were particularly vulnerable to erosion in the South because rainfall intensities could reach up to one and a half inches per hour and the frozen ground and snow cover in the North allowed for little erosion during winter storms. In addition, the South's topography is carved into rougher slopes than the gentle contours of glacier-sculpted New England.

Erosion continued to degrade the South after the Civil War. After surveying regional erosion problems in the southern Appalachians from 1904 to 1907 for the U.S. Geological Survey, Leonidas Chalmers Glenn described farming practices little changed from colonial days.

> When first cleared, the land is usually planted in corn for about two or three years, is then for two or three years put in small grain . . . and then back into corn for several years. Unless it is well cared for the land has by this time become poor, for it has lost its original humus. The soil has become less porous and less able to absorb the rainfall and erosion begins. Means are rarely taken to prevent or check this erosion, so it increases rapidly and the field is soon abandoned and a new one cleared. . . . Many fields are worn out and abandoned before the trees girdled in its clearing have all fallen. Then new grounds are usually cleared beside the abandoned field and the same destructive process is repeated.

It took a few hundred years, but agricultural clearing was finally reaching into the remotest uplands of the region in a process much like what happened in Greece, Italy, and France. "In some places it was found that the entire surface wore away slowly, each heavy rain removing a thin layer or sheet of material, so that the fertile soil layer gradually wore thin and poor and the field was at last abandoned as worn out. . . . Sheet-wash erosion is so slow and gradual that some farmers fail to recognize it and believe that their soils have deteriorated through exhaustion of the fertility, whereas they have slowly and almost imperceptibly worn away to the subsoil."[29]

By the early 1900s more than five million acres of formerly cultivated land in the South lay idle because of the detrimental effects of soil erosion.

When the government began to support aggressive soil conservation efforts in the 1930s, the new U.S. Soil Conservation Service did not offer

Figure 17. Eroded land on tenant farm, Walker County, Alabama, February 1937 (Library of Congress, LC-USF346–025121-D).

up radical new ideas. "Most of the erosion-control practices in use at the present time, such as the use of legumes and grasses, deep plowing, contour plowing and hillside ditching, the prototype of modern terracing, were either developed by the Virginia farmers or became known to them during the first half of the nineteenth century."[30] Actually, most of these techniques, or similar practices, had been used in Europe for centuries and were known in Roman times. If these ideas were so good and had been around for so long why did they take so long to become widely adopted? While Thomas Jefferson and George Washington might disagree on both the reason and the cure, the lessons of the Old World and colonial America remain on the sidelines as a similar story unfolds in the Amazon basin, where the Brazilian government has a long history of encouraging peasants to clear rainforest in order to pacify demands for land reform.

Ironically, the Amazon itself holds clues to a solution. Archaeologists recently discovered areas with incredibly fertile black soil not far from the Carajás Plateau. This rich dirt, called *terra preta,* may cover as much as a tenth of Amazonia. Not only did this distinctly untropical soil sustain large settlements for several thousand years, but intensive habitation produced it. Faced with trying to make a living from nutrient-poor soils, Amazonians improved their soil through intensive composting and soil husbandry.

Found on low hills overlooking rivers, terra preta is full of broken ceramics and organic debris with a high charcoal content and evidence of concentrated nutrient recycling from excrement, organic waste, fish, and animal bones. Abundant burial urns suggest that the human population recycled itself too. The oldest deposits are more than two thousand years old. Practices that built terra preta soils spread upriver over a span of about a thousand years and worked well enough for sedentary people to prosper in an environment that had previously supported a sparse, highly mobile population.

Typically one to two feet thick, deposits of terra preta can reach more than six feet deep. In contrast to the typical slash-and-burn agriculture of tropical regions, Amazonians stirred charcoal into the soil and then used their fields as composting grounds. With almost twice the organic matter as adjacent soils, terra preta better retains nutrients and has more soil microorganisms. Some soil ecologists believe that Amazonians added soil rich in microorganisms to initiate the composting process, as a baker adds yeast to make bread.

Radiocarbon dating of terra preta at Açutuba near the confluence of the Amazon and the Rio Negro showed that the site was occupied for almost two thousand years, from when the black soils began forming about 360 BC until at least AD 1440. When Francisco de Orellana traveled up the Amazon River in 1542 he found large settlements no more than "a crossbow shot" from each other. His conquistadors fled from the throngs of people that swarmed the river at a large site near the mouth of the Tapajós River where terra preta covering several square miles could have supported several hundred thousand people.

Geographer William Denevan argues that slash-and-burn agriculture in which farmers move their plots every two to four years is a relatively recent development in the Amazon. He asserts that the difficulty of clearing huge hardwood trees with stone tools rendered frequent clearing of new fields impractical. Instead, he believes that Amazonians practiced intensive agroforestry that included understory and tree crops that together protected the fields from erosion, allowing rich black earth to build up through time.

Much like a villagewide compost heap, terra preta soils are thought to build up from mixing ash from fires and decomposing garbage into the soil. Similar darkening and enrichment of the soil has been noted around villages in the jungle of northeastern Thailand. Native communities often kept fires burning at all times and terra preta deposits appear lens shaped, suggesting accumulation within, rather than around villages. Relatively high phosphorus and calcium content of terra preta also suggests contributions from ash, fish and animal bones, and urine. Estimated to have grown by an inch in twenty-five years, six feet of terra preta could build up after several thousand years of continuous occupation. Today, terra preta is dug up and sold by the ton to spread on yards in urbanizing parts of Brazil.

Whether catastrophically rapid or drawn out over centuries, accelerated soil erosion devastates human populations that rely on the soil for their living. Everything else—culture, art, and science—depends upon adequate agricultural production. Obscured in prosperous times, such connections become starkly apparent when agriculture falters. Recently, the problem of environmental refugees fleeing the effects of soil erosion began to rival political emigration as the world's foremost humanitarian problem. Although usually portrayed as natural disasters, crop failures and famines often owe as much to land abuse as to natural calamities.

SEVEN

Dust Blow

One man cannot stop the dust from blowing but one man can start it.

FARM SECURITY ADMINISTRATION

NORTHERN CANADA MESMERIZED ME THE FIRST TIME I flew over the pole from Seattle to London on a clear day. While the other passengers enjoyed some Hollywood epic, I drank in the vast plain of bare rock and shallow lakes crawling by six miles below. For tens of millions of years before the onset of the glacial era, deep soil and weathered rock covered northern Canada. Redwood trees grew in the Arctic. Then, as the planet cooled into a glacial deep freeze about two and a half million years ago, rivers of ice began stripping northern Canada down to hard rock, dumping the ancient soil in Iowa, Ohio, and as far south as Missouri. High winds dropping off the great ice sheet blew the pulverized dirt around to shape Kansas, Nebraska, and the Dakotas. Today, these geologic dust bunnies produced by extreme erosion form the best agricultural lands on the planet.

Glaciers also stripped soils from northern Europe and Asia, redistributing thick blankets of finely ground dirt—loess—over more than a fifth of Earth's land surface. Mostly silt with some clay and a little sand, loess forms ideal agricultural soil. Scraped off the Arctic by glaciers and dropped in temperate latitudes by strong winds, the deep loess soils of the world's breadbaskets are incredibly fertile owing to a high proportion of fresh minerals. The absence of stones makes loess relatively easy to plow. But with little natural cohesion, loess erodes rapidly if stripped of vegetation and exposed to wind or rain.

Grazed by buffalo for at least two hundred thousand years, the Great Plains had a thick cover of tough grass that protected the fragile loess. Wandering across the plains the great herds manured the grasslands, enriching the soil. Much of the biomass lay below ground in an extensive network of roots that supported the prairie grass. Traditional plows could not cut through the thick mat that held the plains together. So the first settlers simply kept heading West.

Then in 1838 John Deere and a partner invented a steel plow capable of turning up the prairie's thick turf. When he began selling his unstoppable plow, Deere set the stage for a humanitarian and ecological disaster because, once plowed, the loess of the semiarid plains simply blew away in dry years. Deere sold a thousand of his new plows in 1846. A few years later he was selling ten thousand a year. With a horse or an ox and a Deere plow a farmer could not only plow up the prairie sod, but farm more acreage. Capital began to replace labor as the limiting factor in farm production.

Another new labor-saving machine, Cyrus McCormick's mechanized harvester helped revolutionize farming and reconfigure the relation between American land, labor, and capital. The McCormick reaper consisted of a blade driven back and forth by gears as the contraption cut and stacked wheat while it advanced. McCormick began testing designs in 1831; by the 1860s thousands of his machines were being assembled each year at his Chicago factory. With a Deere plow and a McCormick reaper a farmer could work far more land than his predecessors.

In the early 1800s American farms relied on methods familiar to Roman farmers, broadcasting seed by hand and walking behind plows pulled by horses or mules. The amount of labor available to a typical family limited the size of farms. Early in the twentieth century tractors replaced horses and mules. At the end of the First World War, there were about 85,000 tractors working on U.S. farms. Just two years later the number tripled to almost a quarter of a million. With steel plows and iron horses, a twentieth-century farmer could work fifteen times as much land as his nineteenth-century grandfather. Today, farmers can plow up eighty acres a day listening to the radio in the air-conditioned cab of a leviathan tractor unimaginable to John Deere, let alone a Roman farmer.

As they spread west, Deere's magical plows turned formerly undesirable land into a speculator's paradise. The Territory of Oklahoma (Indian territory, in Chocktaw) was set aside as a reservation for the Cherokee, Chickasaw, Choctaw, Creek, and Seminole nations in 1854. It did not take long before the Indians' practice of maintaining open prairie seemed a waste to land-hungry settlers. From 1878 to 1889 the U.S. Army forcibly ejected white settlers who encroached on Indian lands. Commercial interests and citizens eager to work the rich soil increasingly threatened treaty commitments made to people who just decades before had ceded their ancestral claim to the eastern seaboard in exchange for Oklahoma and the right to be left alone. Eventually bowing to public pressure, the government announced plans to open the territory to settlers in the spring of 1889.

From mid-March into April thousands of people flocked to Oklahoma's borders. Potential settlers were allowed to peruse Indian lands the day before the district opened. The land grab began at noon on April 22 (now celebrated as Earth Day) as the cavalry watched mobs race to stake out their turf. "Sooners" who had slipped by the border guards began filing papers to claim the best land for town sites and farms. By nightfall entire towns were staked out; many homesteads had multiple claimants. Within a week, Indian territory's more than fifty thousand new residents accounted for the majority of its population.

The following year, congressional aid prevented disaster when the settlers' first crops withered. The average rainfall of just ten inches a year could barely support the drought-adapted native grass, let alone crops. In contrast to prairie grass, which weathered dry years and held the rich loess soil in place, a sea of dead crops bared loose soil to high winds and thunderstorm runoff.

Recognizing the potential for an agricultural catastrophe, Grand Canyon explorer and director of the new U.S. Geological Survey Major John Wesley Powell recommended that settlers in the semiarid West be allowed to homestead twenty-five hundred acres of land, but be allotted water to irrigate just twenty acres. He thought this would both prevent overuse of water and conserve the region's fragile soils. Instead, Congress retained the allocation of one hundred and sixty acres of land to each homesteader wherever they settled. That much land could yield a fortune in California. On the plains, an industrious family could starve trying to farm twice as much.

Undeterred by nay-saying pessimists, land boosters advertised the unlimited agricultural potential of the plains, popularizing the notion that "rain follows the plow." It certainly helped their pitch that settlers started plowing the Great Plains during a wet spell. Between 1870 and 1900, American farmers brought as much virgin land into cultivation as they had in the previous two centuries. Mostly crops were good at first. Then the drought came.

The late nineteenth-century advent of widespread lending encouraged Oklahoma's new farmers to borrow liberally and pay off the interest on their loans by mining soil in aggressive production for export markets. Just over two decades after the Oklahoma land rush, farmers plowed up forty million acres of virgin prairie to cash in on high grain prices during the First World War. In the above-average rainfall of the early 1900s, millions of acres of prairie became amber fields of grain. Relatively few paused to consider what would happen should high winds accompany the next inevitable drought.

In 1902 the twenty-second annual report of the U.S. Geological Survey concluded that the semiarid High Plains from Nebraska to Texas were fatally vulnerable to rapid erosion if plowed: "The High Plains, in short, are held by their sod." With rainfall too low to support crops consistently, grazing was the only long-term use for which the "hopelessly nonagricultural" region was well suited.[1] Once stripped of sod, the loess soil would not stay put under the high winds and pounding rains of the open prairie. The survey's findings were no match for land speculation and the high crop prices during the First World War. A century later, talk of returning the region to large-scale grazing as a buffalo commons echoes the survey's far-sighted advice.

Half of the potential farmland in the United States was under cultivation at the end of the nineteenth century. Even conservative textbooks held that static crop yields despite significant technological advances meant soil fertility was declining. Soil erosion was recognized as one of the most fundamental and important resource conservation problems facing the nation. Harvard University geology professor Nathaniel Southgate Shaler even warned that the rapid pace of soil destruction threatened to undermine civilization.

Protecting society's fundamental interest in the soil was not just the government's job, Shaler held, it was one of its primary purposes. "Soil is a kind of placenta that enables living beings to feed on the earth. In it the substances utterly unfit to nourish plants in the state in which they exist in the rocks are brought to the soluble shape whence they may be lifted into life. All this process depends on the adjustment of the rate of rock decay to that of . . . renewing soil." Shaler recognized that agricultural practices mined soil fertility by eroding soil faster than it formed. "The true aim . . . of a conservative agriculture . . . is to bring about and keep the balance between the processes of rock decay and erosion. . . . With rare exceptions, the fields of all countries have been made to bear their crops without the least reference to the interests of future generations."[2] Shaler considered those who abused land to be among the lowest of criminals.

Shaler understood how plowing altered the balance between soil production and erosion. "In its primitive state the soil is each year losing a portion of its nutrient material, but the rate at which the substances go away is generally not more rapid than the downward movement of the layer into the bed rock. . . . But when tillage is introduced, the inevitable tendency of the process is to increase the rate at which the soil is removed."[3] Disturbance of such a balance led to predictable consequences.

Satisfied that modern evidence supported his ideas, Shaler concluded that soil erosion shaped ancient history throughout the Old World. Once the soil was lost, recovery lay beyond history's horizon. "Where subsoil as well as the truly fertile layer has been swept away the field may be regarded as lost to the uses of man, as much so, indeed, as if it had been sunk beneath the sea, for it will in most instances require thousands of years before the surface can be restored to its original estate."[4] The six thousand square miles of fields by then abandoned to erosion in Virginia, Tennessee, and Kentucky testified to the American tendency to repeat Old World mistakes.

Although Shaler advised tilling down through the subsoil to break up decaying bedrock and speed soil creation, he argued that land sloping more than five degrees should be spared the plow. He predicted that fertilizers could replace rock weathering, but did not foresee how mechanized agriculture would further increase erosion rates on America's farmlands.

Nonetheless, soil erosion was becoming a national problem. In 1909 the National Conservation Congress reported that almost eleven million acres of American farmland had been abandoned because of erosion damage. Four years later, the U.S. Department of Agriculture (USDA) estimated that annual topsoil loss from U.S. fields amounted to more than twice the quantity of earth moved to dig the Panama Canal. Three years after that, Agricultural Experiment Station researchers estimated that half the tillable land in Wisconsin suffered from soil erosion that adversely affected economic activity.

At the start of the First World War, the USDA's annual yearbook lamented the economic waste from soil erosion. Rain fell like "thousands of little hammers beating upon the soil" and ran off bare ground in rivulets that were slowly stealing the nation's future. "Under the original process of nature the soil was continually wearing away on the top, but more was forming, and the formation was somewhat more rapid than the removal. The layer of soil on hillsides represented the difference between the amount formed and that removed. After clearing, the rate of removal is greatly increased, but the rate of formation remains the same."[5] Already more than three million acres of farmland had been ruined by erosion. Another eight million acres were too degraded to farm profitably.

Reclamation of all but the most severely damaged farmland was possible, and potentially even profitable, but it required new farming practices and attitudes.

> Many farmers when approached on the subject of erosion show interest and agree that the loss is great. They will say, "Why, yes, some of my fields are badly washed, but it doesn't pay to try to do anything with them." They expect reclamation, if it is ever accomplished, to be undertaken by the Government, and it is only with difficulty that they can be induced to make an attempt at stopping the ravages of erosion. It has been cheaper in the past to move to newer lands.[6]

Soil loss occurred slowly enough that farmers saw the problem as someone else's concern. Besides, mechanization made it even easier to just plow more land than to worry about soil loss. Machines were expensive and needed to pay for themselves—dirt was cheap enough to ignore losing a little here and there, or even everywhere.

The wide-open plains presented an ideal place for tractors. The first locomotive-like tractors arrived around 1900. By 1917 hundreds of companies were cranking out smaller, more practical models. Before abandoning the market to agricultural specialists like International Harvester and John Deere, Henry Ford invented a rear hitch that allowed tractors to pull plows, disks, scrapers and other earth-moving equipment across farms. Armed with these marvelous machines, a farmer could work far more land than he had when trailing behind an ox or horse. He could also plow up the pasture to plant more crops.

The cost of the new machines added up to more than many small farms could afford. From 1910 to 1920 the value of farm implements on a typical Kansas farm tripled. In the next decade costs tripled again as more farmers bought more tractors, trucks, and combines. When prices for grain were high, it was profitable to operate the machinery. When prices dropped, as they did after the First World War, many farmers were saddled with unmanageable debt. Farmers who stayed in business saw bigger machines working more land as the way to a secure future. Just as the English land enclosures of the seventeenth and eighteenth centuries had displaced poor

Figure 18. Breaking new land with disk plows, Greely County, Kansas, 1925 (courtesy of Kansas State Historical Society).

peasants, the spread of tractors displaced those lacking the capital to join the party.

By 1928, when Hugh Bennett and W. R. Chapline published the first national soil erosion assessment, topsoil loss amounted to five billion tons a year—several times faster than soil loss in the nineteenth century and ten times faster than soil formed. Nationwide, virtually all the topsoil had already eroded from enough farmland to cover South Carolina. Six years later Bennett and Chapline's report seemed understated. Even in drought and Depression, the number of tractors working Oklahoma farms increased from 1929 to 1936. New disk plows, with rows of concave plates set out along a beam, thoroughly diced the upper layers of soil, leaving a pulverized layer that could easily blow away in dry conditions.

The first major windstorm of 1933 swept through South Dakota on November 11. Some farms lost all their topsoil in a single day. The next morning the sky remained dark until noon—one part air to three parts dust. No one knew this was just a preview.

On May 9, 1934, fields from Montana and Wyoming were ripped up by high winds. Blowing across the Dakotas, the wind kept picking up dirt until a third of a billion tons of topsoil was heading east at up to a hundred miles an hour. In Chicago four pounds of dust dropped out of the sky for each person in the city. The next day Buffalo, in eastern upstate New York, fell dark at noon. By dawn on May 11 dust was settling on New York City, Boston, and Washington. The huge brown cloud could be seen far out in the Atlantic Ocean.

Resilient when under permanent vegetation, grazed (and manured) by millions of buffalo, the prairie fell apart when plowed up and dried out by prolonged drought. Without the grasses and their roots that stabilized the soil, winds that decades before blew harmlessly across the range ripped the countryside open like a sand-charged hurricane. Shifting drifts of dirt covered a vast region where high winds carried off desiccated soil exposed beneath the parched stubble of wilted crops. High winds stirred up enough dust to choke people, shred crops, kill livestock, and shroud distant New York City in an eerie veil.

The National Resources Board reported that by the end of 1934, dust storms had destroyed an area larger than the state of Virginia. Another hundred million acres were severely degraded.

In the spring of 1935 strong winds again tore through the parched fields of Kansas, Texas, Colorado, Oklahoma, and Nebraska. With the fields freshly plowed, there was no vegetation to hold the dry loess in place. The finest and most fertile soil formed dark blizzards rising ten thousand feet to blot out the midday sun. Coarser sand blew around near the ground, gnawing through fence posts. Streetlights stayed on all day. High winds piled up Sahara-like dunes, blocking trains and paralyzing the plains.

On April 2, 1935, Hugh Bennett testified before the Senate Public Lands Committee about the need for a national soil conservation program. Bennett knew that a great dust storm from the plains was descending on Washington. With help from field agents who called to report the progress of the dirt cloud, he timed his testimony so that the sky went dark as he presented it. Duly impressed, Congress appointed Bennett head of a new Soil Conservation Service.

The agency faced a daunting challenge. Within a few decades of settlement, barren desert had replaced the short-grass prairie. President Franklin D. Roosevelt ended the era of land settlement in November 1934 by closing the remaining public lands to homesteading. American agricultural expansion was officially over. Displaced Dust Bowl farmers had to find work in someone else's fields.

More than three million people left the plains in the 1930s. Not all of them were fleeing dust, but about three-quarters of a million displaced farmers headed west. The grandchildren of the original sooners became

Figure 19. Dust storm approaching Stratford, Texas, April 18, 1935 (NOAA, George E. Marsh Album; available at www.photolib.noaa.gov/historic/c&gs/theb1365.htm).

environmental refugees, unwelcome until they reached California's new, labor-hungry fields at the edge of the continent.

The problem of soil erosion was not restricted to the Dust Bowl. In 1935 the Department of Agriculture estimated the amount of ruined and abandoned farmland at up to fifty million acres. Two to three times that much land was losing an inch of topsoil every four to twenty years. Two hundred thousand acres of abandoned Iowa farmland was eroded beyond redemption. The next year the new Soil Conservation Service reported that more than three-quarters of Missouri had lost at least a quarter of its original topsoil, more than twenty billion tons of dirt since the state was first cultivated. Only four of their original sixteen inches of topsoil remained on some fields. The U.S. Bureau of Agricultural Engineering reported that it was common for southeastern farms to lose more than six inches of soil in a single generation. In the aftermath of the Dust Bowl, which cost over a billion dollars in federal relief, the federal government began to see soil conservation as an issue of national survival.

State and federal commissions traced the severity of the 1930s dust storms to a tremendously increased acreage under cultivation, much of which was

Figure 20. Buried machinery in barn lot, Dallas, South Dakota, May 13, 1936 (USDA image No:00di0971 CD8151–971; available at www.usda.gov/oc/photo/00di0971.htm).

marginal land. The Kansas State Board of Agriculture, for example, blamed the disaster on poor farming practices. "Soil has been cultivated when extremely dry, and no effort has been made, in most cases, to return organic matter to the soil. . . . When cultivated in a dry condition such a soil became loose and dusty. There are individual farmers throughout the region who have followed good methods of soil management and have found it possible to prevent soil blowing on their farms, except where soil blown from adjoining farms encroached upon their fields."[7] The report of the Great Plains Committee convened in 1936 by the House of Representatives had identified economic forces as a major cause of the disaster.

> The [First] World War and the following inflation pushed the price
> of wheat to new high levels and caused a remarkable extension of the
> area planted to this crop. When the price collapsed during the post-war
> period Great Plains farmers continued to plant large wheat acreages
> in a desperate endeavor to get money with which to pay debt charges,

taxes, and other unavoidable expenses. They had no choice in the matter. Without money they could not remain solvent or continue to farm. Yet to get money they were obliged to extend farming practices which were collectively ruinous.[8]

Walter Lowdermilk, by then the associate chief of the Soil Conservation Service, suggested using the erosion rate on undisturbed lands as the geologic norm of erosion to provide the benchmark for gauging human-induced erosion. His concern seemed justified when the Soil Conservation Service compiled county-level soil erosion maps into a national map. The results were alarming. More than three-quarters of the original topsoil had been stripped off almost two hundred million acres of land, about a tenth of the area surveyed. From one- to three-fourths of the topsoil was gone from two-thirds of a billion acres, more than a third of the area surveyed. At least a quarter of the soil was missing from almost a billion acres of land. America was losing its dirt.

In a speech given before the annual meeting of the National Education Association in July 1940, Hugh Bennett would describe the dust storm of May six years earlier as a turning point in public awareness. "I suspect that when people along the seaboard of the eastern United States began to taste fresh soil from the plains 2,000 miles away, many of them realized for the first time that somewhere something had gone wrong with the land."[9]

On April 27, 1935, Congress had declared soil erosion a national menace and established the Soil Conservation Service to consolidate federal actions under a single agency. A year later in his opening address to a conference convened by order of President Roosevelt, the agency's newly appointed chief Hugh Bennett compared the rapid loss of soil from U.S. farmlands to the slow pace of soil formation.

Citing federal studies, Bennett showed just how fast America was disappearing. The erosion research station at Tyler, Texas found that the region's best farming practices increased soil loss by almost two hundred times the soil replacement rate. Poor management practices increased erosion by eight hundred times. The research station at Bethany, Missouri showed that soil loss from typical corn lands was three hundred times that of comparable land under alfalfa.

Research also showed that after erosion of the loose topsoil, more rainfall ran off over the surface instead of sinking into the ground. This produced more runoff, which then removed even more soil, producing yet more runoff. It did not take long to lose topsoil once the process started.

Bennett calculated that it took more than five thousand years for rainfall to remove six inches of topsoil from native grassland in Ohio. This made sense; it was close to the rate at which he thought soil formed—about an inch every thousand years. In contrast, fields lost some six inches of topsoil in little more than three decades of continuous cultivation. The erosion research station at Guthrie, Oklahoma found that the fine sandy loam covering the plains eroded more than ten thousand times faster under cotton cultivation than native grass. Cotton farming could strip off the region's typical seven inches of topsoil in less than fifty years. The same topsoil under grazed grass would last more than a quarter million years. The message was clear, Bennett advised not plowing hilly and highly erodible land.

Echoing Bennett's warnings in 1953, his associate chief described how almost three-quarters of U.S. farmland was losing soil faster than it formed. In particular, Lowdermilk stressed that the United States was following ancient civilizations down the road to ruin. He argued that seven thousand years of history cautioned against plowing hillslopes.

> Here in a nutshell, so to speak, we have the underlying hazard of civilization. By clearing and cultivating sloping lands—for most of our lands are more or less sloping—we expose soils to accelerated erosion by water or by wind. . . . In doing this we enter upon a regime of self-destructive agriculture. . . . As our population increases, farm production will go down from depletion of soil resources unless measures of soil conservation are put into effect throughout the land.[10]

Lowdermilk did not see this as a remote threat some centuries in the future. He viewed the wars of the twentieth century as a battle for control of land.

After the Second World War, conversion of military assembly lines to civilian uses dramatically increased tractor production, completing the mechanization of American farms and paving the way for high-output industrial farming in developed countries. Several million tractors were working American fields by the 1950s—ten times as many as in the 1920s. The number of U.S. farmers plummeted as farm acreage increased and more people moved to the swelling cities. The few farmers left on their land grew cash crops to pay off the loans on their new labor-saving equipment. Mechanization, like slave labor in the South, required doing the same thing everywhere instead of adapting agricultural methods to the land.

Figure 21. Plowing a steep hillside circa 1935 (National Archives, photo RG-083-G-36711).

Droughts in the Great Plains occur about every twenty years. In the wet 1940s, doubling the acreage under cultivation increased wheat production fourfold—enough to support record exports to Europe during the war. In 1956 drought again caused near failure of the wheat crop. The 1950s drought lasted almost as long as the 1930s drought and was as severe as the 1890s drought (though this time soil conservation programs were widely credited with preventing another Dust Bowl). Small farms were going bankrupt while large farms better able to withstand the periodic dry spells bought more and larger machinery.

The U.S. government had begun farm subsidies in 1933. Within a year, most Great Plains farmers were participating in programs aimed at soil conservation, crop diversification, stabilizing farm income, and creating flexible farm credit. As much as anything, this last element, which allowed farmers to carry more debt, changed American farming. Within a decade, farm debt more than doubled while farm income rose by just a third. Despite a continual rise in government subsidies, more than four out of every ten American farms disappeared between 1933 and 1968. Corporate factory farms better able to finance increasingly expensive farm machinery and agrochemicals began to dominate American agriculture by the end of the 1960s.

Although different in detail from Rome and the South, the economics of large corporate farms similarly discounted concern about soil erosion.

> Corporations are, by nature, temporary land owners. . . . A tenant on corporate land has no assurance whatsoever of staying on the farm more than a year. . . . A high proportion of corporate land tends to cause instability in land tenure and to foster erosion, unless the major-ity of corporations can be induced to adopt definite soil conservation programs on their land. Heavy mortgage indebtedness exerts a specific financial pressure upon the soil by forcing the farmer to squeeze out of his soil whatever he can to meet his financial obligations.[11]

The growth of mechanized industrial agriculture promoted rapid soil loss as farmers spent their natural capital to service loans for machinery and fertilizers.

Records at Woburn Experimental Farm, established about twenty-five miles north of London in 1876 by England's Royal Agricultural Society, inadvertently documented the effects of changing agricultural practices on soil erosion. The first half century of crop yield experiments recorded lit-tle erosion. After the Second World War, the introduction of herbicides and heavy farm machinery changed that.

The first report of soil erosion problems came after a storm on May 21, 1950, when intense rainfall carved four-inch-deep, three-foot-wide gullies into bare fields, burying sugar beet plots beneath piles of dirt and unearthed potatoes. Serious erosion in the 1960s sharply reduced the organic nitrogen content on experimental plots. By the 1980s the farm served to validate soil erosion models, as more than a dozen erosional events occurred each year, especially on the farm's steepest slopes. Yet the detailed diaries kept by farm staff from 1882 to 1947 had focused on the subtleties of crop performance, cultivation techniques, soil pH, and crop damage from varmints, with no mention of erosion before the introduction of heavy machinery and agrochemicals. Adopting twentieth-century agricultural methods greatly accelerated soil erosion.

One of the most persistent agricultural myths is that larger mechanized farms are more efficient and profitable than smaller traditional farms. But larger farms spend more per unit of production because they buy expensive equipment, fertilizer, and pesticides. Unlike industrial enterprises in which economies of scale characterize manufacturing, smaller farms can be more efficient—even before accounting for health, environmental, and social costs. A 1989 National Research Council study flatly contradicted the bigger is more efficient myth of American agriculture. "Well-managed alternative farming systems nearly always use less synthetic chemical pesticides, fertilizers, and antibiotics per unit of production than conventional farms. Reduced use of these inputs lowers production costs and lessens agriculture's potential for adverse environmental and health effects without decreasing—and in some cases increasing—per acre crop yields."[12]

Small farms also can produce more food from the same amount of land. A 1992 U.S. agricultural census report found that small farms grow two to ten times as much per acre as do large farms. When compared to farms greater than six thousand acres in size, farms smaller than twenty-seven acres were more than ten times as productive; some tiny farms—less than four acres—were more than a hundred times as productive. The World Bank now encourages small farms to increase agricultural productivity in developing nations, where most landholders own less than ten acres.

A key difference between small farms and large industrial farming operations is that large farms typically practice monoculture, even though they may grow different crops in different fields. Single-crop fields are ideal for heavy machinery and intensive chemical use. Although monocultures generally produce the greatest yields per acre for a single crop, diversified polycultures produce more food per acre based on the total output from several crops.

Despite the overall efficiency of small farms, the trend is toward larger, more industrialized farms. In the 1930s seven million Americans farmed. Today fewer than two million farmers remain on their land. As recently as the early 1990s, the United States had lost more than twenty-five thousand family farms a year. On average, more than two hundred American farms have gone under every day for the past fifty years. In the second half of the twentieth century, the average farm size more than doubled, from under one hundred to almost two hundred hectares. Less than 20 percent of U.S. farms now produce almost 90 percent of the food grown in America.

As crop yields increased two- to threefold from 1950 to the 1990s, the cost of machinery, fertilizer, and pesticide rose from about half to over three-quarters of farm income. Two types of farms survived: those that opted out of industrialization and those that grew by working larger areas for a smaller net return per acre. By the 1980s the largest farms, dubbed superfarms by the USDA, accounted for close to half of all farm income.

If small-scale agriculture is so efficient, why are America's small farms going under? The high capital costs of mechanization can be an economic disaster for a small operation. A farm must be large to profitably use technology-intensive methods instead of labor-intensive methods. Sold on the idea that modernizing meant mechanizing, small farms sank into debt once overleveraged; large companies then bought up their land. This process may not help small farms stay in family hands, but it pumps a lot of cash into companies that produce farm equipment and supplies—and advise farmers how to use their products.

The economic and social trends that drove mechanization turned farming into an industry and accelerated soil loss. New equipment made more intensive cultivation of land easier, to a deeper depth and more often. Just as in ancient Rome, the ground lay bare and disturbed for much of the year. As farms mechanized, soil conservation practices such as terracing, hedgerows, and trees planted for windbreaks became obstacles to maneuvering heavy machinery. Contour plowing practices were modified to accommodate large machines that could not follow tight turns on sloping land. Soil was now a commodity—the cheapest of many inputs to agricultural manufacturing.

Substantial progress in raising both public and governmental awareness slowed but did not stop soil loss. Some areas have fared worse than others. Across the heart of the Midwest, islands of native prairie rising up to six feet above neighboring plowed fields testify to soil loss of about half an inch per year since settlement. Iowa lost half its topsoil in the last century and a half. Fortunate by comparison, the Palouse region of eastern Washington lost only a third to half of its rich topsoil in the past century.

The first settlers arrived in the Palouse in the summer of 1869. They grew grain on the valley bottoms and raised cattle and hogs to sell to miners in nearby Idaho. The region's deep loess soil could produce more but there was no way to get crops to market. Completion of the railroads in the 1880s opened the land to distant markets, new equipment, and more farmers. By the 1890s most of the Palouse was under cultivation.

Soil erosion rapidly became a major problem once the loess was cleared and plowed. In the early 1900s Washington State Agricultural College's William Spillman toured the region lecturing on the threat of soil erosion from the common practice of leaving plowed fields bare each summer. Few heeded the young professor's warning that each year's annoying rills would eventually add up to a serious problem.

In the 1930s tractors began replacing horse-drawn plows in the Palouse and elsewhere, allowing a single operator to farm much larger acreages. Eager to capitalize on the greater labor efficiency, landowners changed the traditional arrangement for sharecropping on leased land. Instead of keeping two-thirds of what they grew, tenants were now allowed to keep just over half. So tenants worked the land that much harder, reducing outlays for luxuries like erosion control. Farmers were now working more land, but not necessarily making more money.

By 1950 a USDA survey reported that all of the original topsoil was missing from 10 percent of Palouse farmland. Between 25 and 75 percent of the topsoil was missing from an additional 80 percent of the land. Just 10 percent of the region retained more than 75 percent of its original soil. Annual surveys of soil loss from 1939 to 1960 showed an average loss of half an inch a decade. On slopes steeper than about fifteen degrees, soil loss averaged an inch every five years.

A cistern installed in 1911 on a farm near Thornton dramatically illustrates the effect of plowing sloping fields. Originally projecting about a foot and a half above the adjacent hilltop, by 1942 it stuck out nearly four feet above the surrounding field. By 1959 the same cistern stood six feet above the field. Four and a half feet of soil had been plowed off the slope in less than fifty years—about an inch a year. Some eastern Idaho soils that were more than a foot thick in the early twentieth century were barely deep enough to plow by the 1960s when just half a foot of soil remained above bedrock.

From 1939 to 1979 the total erosion on Palouse cropland averaged more than nine tons per acre per year; and reached more than one hundred tons per acre per year on steep slopes. Erosion rates on unplowed rangeland and forested land averaged less than one ton per acre per year. Plowing the loess increased erosion rates by a factor of ten to a hundred, with most of the loss caused by erosion from runoff across newly plowed ground. Simple soil conservation measures could halve erosion without reducing farm income. But doing so requires fundamental changes in farming practices.

Figure 22. Bare, rilled field in the Palouse region of eastern Washington in the 1970s (USDA 1979, 6).

In 1979 the Soil Conservation Service reported that three decades of plowing had lowered fields as much as three feet below unplowed grassland. Berms of soil four to ten feet high stood at the downhill end of plowed fields. Experiments conducted with a typical sixteen-inch moldboard plow pulled along contours showed that plowing typically pushed soil more than a foot downhill. The process that stripped Greek hillsides in the Bronze Age was being repeated in the Palouse.

Simply plowing the land pushed soil downhill far faster than natural processes ever managed. Even so, this process is almost as hard to notice, occurring imperceptibly with each pass of a plow. Continued for generations, till-based agriculture will strip soil right off the land as it did in ancient Europe and the Middle East. With current agricultural technology, though, we can do it a lot faster.

Wind erosion contributes to the problem. A core from the bed of eastern Washington's Fourth of July Lake records that dust fall into the lake increased fourfold with the introduction of modern agriculture to the region. There are few reliable measurements of wind erosion under natural conditions, but under the right conditions it can be extreme. Before soil conservation measures were adopted, wind stripped up to four inches of soil a year from some Kansas fields in the Dust Bowl era. Dust blowing off of bare, dry field is still a problem in eastern Washington. In September 1999 dust blowing off of agricultural fields blinded drivers and triggered fatal traffic accidents on Interstate 84 near Pendleton, Oregon.

Plowing exposes bare, disrupted soil to dramatic erosion when storms ravage ground not yet shielded by vegetation. In the American Midwest, over half the erosion from land planted in corn occurs in May and June before crops grow large enough to cover the ground.

Crop yields fall once the topsoil is gone and farmers plow down into subsoil with lower organic matter, nutrient content, and water-retention capacity. Losing six inches of topsoil from Georgia and west Tennessee soils reduced crop yields by almost half. Severely eroded areas of Kentucky, Illinois, Indiana, and Michigan already produce a quarter less corn than they once did. Just a foot or two of erosion can dramatically reduce soil productivity—sometimes to the point of losing all agricultural potential. Less than 50 percent of U.S. croplands have slopes gentler than 2 percent and therefore little threat of accelerated erosion. The steepest 33 percent of U.S. cropland is projected to fall out of production over the next century. Since 1985 the Grassland Reserve Program has been paying farmers to restore and preserve grasslands in areas vulnerable to soil erosion.

Soil erosion is not only a problem of capitalist agriculture. The rich black earth of the Russian steppe eroded rapidly once the native vegetation was cleared. Although deep gullies surrounded Russian settlements as early as the sixteenth century, the fragile nature of these soils did not slow efforts to industrialize Soviet agriculture in the twentieth century. The first five-year plan, produced in 1929, included a blunt call to convert the steppe to factory farms. "Our steppe will truly become ours only when we come with columns of tractors and ploughs to break the thousand-year-old virgin soil."[13] Contrary to plan, dust storms blossomed after plows broke up the grassland.

The Soviet's virgin land program of the 1950s and 1960s brought a hundred million acres of marginal farmland into production. Against the advice of prominent scientists aware of the American Dust Bowl, Premier Nikita Khrushchev ordered state collectives to plow forty million acres of virgin land from 1954 to 1965. Food production was not keeping pace with postwar consumer demand.

With the ground left bare during fallow periods, severe erosion reduced crop yields within a few years on much of the newly cleared land. At the program's peak, Soviet agriculture lost more than three million acres a year—not a good way to fulfill a five-year plan. Severe erosion damaged almost half the newly plowed land during the next dry spell in the 1960s, creating a little-publicized Soviet dust bowl that helped drive Khrushchev from office.

Before 1986 Soviet censors hid the extent of environmental problems. Foremost among those problems was the Aral Sea disaster. In 1950 the Soviet government initiated a major effort to achieve "cotton independence" by turning the region into monocultural plantations. The Soviets greatly increased crop yields through improved cultivation techniques, aggressive fertilizer and pesticide use, and by expanding irrigation and mechanized agriculture. From 1960 to 1990 thousands of miles of new canals and more than six hundred dams diverted rivers from the Aral Sea. Not surprisingly, the sea began to shrink.

As the Aral Sea dried out, so did the surrounding land. By 1993, decades of continuous water diversion lowered the water level almost fifty-five feet, creating a new desert on the exposed seabed. Major dust storms in the 1990s dropped a hundred million tons of Aral salt and silt on Russian farms a thousand miles away. The collapse of both the fishing industry and agriculture triggered a mass exodus.

A post-glasnost regional assessment revealed that desertification affected two-thirds of arid lands in Kazakhstan, Uzbekistan, and Turkmenistan. Proposals to address this growing threat went nowhere before the Soviet Union disintegrated. Independence only increased the desire to pursue cash crops for export, moving the fight against soil erosion to the bottom of the political agenda. Despite the clear long-term threat, more immediate concerns prevailed.

A similar situation developed in the small southern Russian Kalmyk Republic tucked between the River Volga and the Caspian Sea. Between the Second World War and the 1990s, aggressive plowing of rangelands desertified most of the republic. Almost a tenth of the country turned into barren wasteland.

Kalmykia's native grasslands were ideal for livestock. As early as the twelfth century, Kalmyks brought cattle to the region where horses were said to graze without bending their heads. Traditional land use centered around horse breeding and sheep or cattle grazing. Accused of collaborating with the Germans, Kalmyks were exiled en masse to Siberia in 1943. By the time they returned fifteen years later, the Soviets were busy creating Europe's first desert.

Throughout the cold war, Soviet policies favored plowing up Kalmyk pastures to increase cereal and melon production. The number of sheep almost doubled on the remaining grassland. Forage crop yields declined by half between the 1960s and the 1990s. Each year the desert consumed another 50,000 hectares of bare fields and overgrazed pasture. By the 1970s more than a third of the republic was partially desertified.

Plowing the native grasslands of this semiarid region led to problems reminiscent of the Dust Bowl. Developed on sand deposited on the bed of the once more extensive Caspian Sea, Kalmykia's rich soils were held together by the roots of the lush native grass. Within decades of plowing, more than a third of a million hectares of grassland were transformed into moving sand seas. In 1969, after extensive agricultural development, a major dust storm blew soil to Poland. Fifteen years later another dust storm sent Kalmyk dirt all the way to France. The republic's president pronounced a state of ecological emergency on August 1, 1993—the first such proclamation from a national government in regard to soil erosion.

The superpowers of the late twentieth century were not alone in losing soil faster than nature makes it. Erosion outpaces soil production by ten to twenty times in Europe. By the mid-1980s roughly half of Australia's agricultural soils were degraded by erosion. Soil erosion from steep slopes in the Philippines and Jamaica can reach four hundred tons per hectare per year—the equivalent of carting off almost an inch and a half of soil a year. Half of Turkey is affected by serious topsoil erosion. Once done, the damage lasts for generations.

In the 1970s sub-Saharan Africa experienced its own dust bowl. Until the twentieth century, West African farmers used a shifting pattern of cultivation that left fields fallow for long periods. Grazing was light as animal herders moved long distances across the landscape each year. In the twentieth century, the combination of a rising population and encroachment of agricultural fields on traditional pastures intensified land use by both farmers and pastoralists. Extensive land clearing and degradation led to extreme soil loss that created a flood of environmental refugees.

The African Sahel lies in the semiarid zone between equatorial forests and the Sahara. On average the region receives six to twenty inches of rainfall annually. But the rain varies widely year to year. In a good year, it rains more than one hundred days in northern Senegal; in bad years it rains fewer than fifty. Studies of ancient lake levels show that long droughts occurred repeatedly over the past several thousand years. Tree ring studies from the Atlas Mountains just north of the Sahel reveal at least six droughts lasting twenty to fifty years between AD 1100 and 1850. The next run of dry years proved catastrophic after almost a half million square miles of West African forest were cleared in under a century.

The 1973 West African famine killed more than a hundred thousand people and left seven million dependent on donated food. Triggered by drought, the roots of the crisis lay in the changing relationship of the people to the land. Extensive removal of ground-protecting vegetation triggered severe soil erosion and a humanitarian disaster during the next run of drier than average years.

The nomads and sedentary farmers of the Sahel traditionally practiced a symbiotic arrangement in which the nomads' cattle would graze on crop stubble, manuring farmers' fields after the harvest. When the rains came the herds would head north following the growth of new grass. Continuing north until the grass was no longer greener ahead of them, the nomads would turn back south, their cattle grazing on the grass that grew up behind them after they passed through on their way north. They would arrive back in the south in time to graze and manure the farmers' harvested fields. In addition, Sahelian farmers grew a variety of crops and let land lie fallow for decades between episodes of cultivation. The division of the Sahel into separate states disrupted this arrangement.

Rapid expansion of French colonial authority across the Sahel in the late nineteenth century altered the social conventions that had prevented overgrazing and kept fields manured. Colonial authorities set up merchants in new administrative centers to stimulate material wants. Poll and animal taxes compelled subsistence farmers and nomads alike to produce goods for French markets. Held to new political boundaries, nomadic tribesmen who had moved their herds across the landscape for centuries increased their livestock density to pay taxes. Farmers moved north into marginal lands to plant crops for export to Europe. Pastoralists expanded south into areas where lack of reliable water and insecurity had previously limited the number of cattle and sheep. Large concentrations of animals around new wells destroyed pastures and left the soil vulnerable to erosive runoff and high winds during violent summer storms.

The Sahel became more evenly and continuously used for increasingly intensive grazing and farming. Between 1930 and 1970, the number of grazing animals doubled; the human population tripled. New French plantations to grow cotton and peanuts as cash crops pushed subsistence farmers onto smaller areas of marginal land. Fallow periods were reduced or eliminated and crop yields began to fall. Ground exposed beneath parched crops dried out and blew away with the wind.

Then in 1972, no rain fell and no grass grew. Livestock mortality was high where continuous overgrazing left little grass from the previous year. The few fruit trees that survived bore little fruit. Millions of refugees flooded into huge shantytowns. Between a hundred thousand and a quarter of a million people died of hunger. While the drought was the immediate cause of this disaster, colonial-era cultural and economic changes led to exploitation of the Sahel and allowed the population to grow beyond what the land could sustain during dry spells. It didn't help that crops grown on large plantations were still exported during the famine.

Destruction of perennial plant cover from overgrazing causes desertification by exposing the soil surface to erosion from wind and rain. Erosion rates of half an inch to three-quarters of an inch per year have been reported in semiarid regions once the native perennials are gone. The process is generally irreversible; plants cannot survive the dry season without the water-holding topsoil. Once the soil is gone, the ability to support people disappears too.

During the famine, an image taken by a NASA satellite provided stark confirmation of the human hand in creating the crisis. A mysterious green pentagon in the center of the drought-ravaged zone turned out to be a quarter-million-acre ranch separated from the surrounding desert by a simple barbed wire fence. The ranch, established in the same year that the drought began, was divided into five sectors, with the cattle allowed to graze one sector each year. Limiting the intensity of grazing prevented the problems that brought starvation to the surrounding countryside.

Desertification began in both the Sahel and North Africa during the 1950s and 1960s, despite above average North African rainfall during these years. Large state-owned ranches established during the 1960s showed no evidence of desertification if stocked at the estimated long-term capacity of the rangelands. Although drought reinforces the effects of land degradation, climate variability is not the root cause. Droughts naturally recur in semiarid regions. Drought-adapted ecosystems and societies weathered them in the past. Traditional African pastoralists practiced de facto population control through social structures and rules developed over centuries of alternating scarcity during droughts and abundance during wetter times.

Soil erosion rates from West African fields range from about three-quarters of an inch per century on savanna cropland to an extreme of more than ten inches a year on bare-plowed fields in steep formerly forested areas. Some estimates put average erosion rates for Sahel cropland at about an inch per year. In many parts of West Africa the topsoil is only six to eight inches thick. Cultivation after forest clearing quickly strips it off. Maize and cowpea yields in southwestern Nigeria dropped by 30 to 90 percent with the loss of less than five inches of topsoil. As the Nigerian population increased, subsistence farmers moved to steeper land that could not support sustained cultivation. Cassava plantations on land with slopes steeper than eight degrees lose soil more than seventy times faster than fields sloping less than a degree. Soil erosion rates on Nigerian hillslopes planted in cassava reach over an inch a year, far beyond any conceivable replacement rate.

Social conventions hindered soil conservation. Subsistence farmers were reluctant to invest in erosion control because they moved their fields every few years. Erosion problems were most severe in areas where communal land ownership discouraged individual efforts to conserve soil. In many West African countries, tractor-hiring schemes are heavily subsidized, so farmers get their fields plowed regardless of the steepness, soil type, or cropping system. Soil erosion rates in sub-Saharan Africa increased twentyfold in the past thirty years. The rapid soil erosion typical of West African agriculture means that only a few years of cultivation are needed to ruin the soil. This, in turn, fuels the drive to clear more land.

In the late 1970s University of Washington professor Tom Dunne and two of his graduate students—one of whom was my graduate advisor—compared recent and long-term erosion rates on the gentle slopes of semi-arid grazing lands in Kenya by using the height of dirt pedestals where vegetation of known (or reasonably estimated) age still held the soil on denuded slopes as well as the amount of incision into land surfaces of known geologic ages. Remnant mounds of soil standing eight inches above the general ground surface around the base of fifteen- to thirty-year-old dwarf shrubs indicated modern erosion rates on the order of a quarter to half an inch per year.

Dunne's team determined that the average rate of erosion since the time of the dinosaurs averaged about an inch every three thousand years. The average erosion rate over the past several million years was about an inch every nine hundred years, a little higher than their estimated rate of soil formation of no more than about an inch every two thousand five hundred years. Present rates of erosion, however, ranged from about an inch every decade to half an inch a year. Based on the discrepancy between rates of soil formation and modern erosion rates, they estimated that it would take between two and ten centuries to strip Kenya's gentle slopes to bare rock.

Soil erosion can destroy the vitality of the land—but land can be healed too. Some subsistence farmers in Nigeria made a few simple changes and transformed their fields—at no cost. Tethering their sheep and feeding them crop stubble instead of letting them wander freely allowed collecting manure to fertilize the next crop. Planting cowpeas as part of a crop rotation also helped enhance soil fertility. Low earth-and-stone walls built around the fields kept soil from leaving in heavy rainfall. Crop yields doubled or even tripled without chemical fertilizer. What was required was labor—exactly what subsistence farmers can afford to give. Labor-intensive techniques that restore soil fertility turned the liability of a dense population into an asset.

Ethiopia provides another example of how human societies more often bring soil erosion with them. Medieval deforestation of the northern portion of the kingdom triggered such extensive erosion in Tigre and Eritrea that the hillsides could no longer support grazing animals. By about AD 1000 the economic impact of soil degradation forced the kingdom to relocate its capital to better land in the south. There the process was repeated, as extensive soil erosion followed widespread deforestation. The region remains impoverished, unable to feed itself when the weather doesn't cooperate.

Drought-triggered crop failure brought starvation to almost ten million people in Ethiopia in the mid-1980s. Hundreds of thousands died despite the largest global famine relief effort in history. Long before the twentieth century, farming had expanded from the best agricultural lands onto erosion-prone slopes. Since the 1930s, deforestation left Ethiopia with just 3 percent of its original forest cover and increased the silt concentration in the Blue Nile fivefold. The average rate of cropland soil loss in the western highlands would erode the native topsoil in little more than a century. In addition to direct losses to erosion, soil fertility has been projected to fall by as much as 1 percent annually owing to persistently intensive cultivation by desperate farmers.

Ethiopia's environmental refugee crisis shows how, over the long run, soil security is national security. Recognition of Wangari Maathai with the 2004 Nobel Peace Prize for her work on environmental restoration in Ethiopia's countryside shows that environmental refugees, who now outnumber political refugees, are an emerging global concern. People may endure temporary droughts, but desertification forces emigration once the land can no longer sustain either grazing or farming.

Desertification is not just happening in Africa. More than a tenth of Earth's land area is desertifying—about a third of the planet's dry lands. Studies over the past fifty years report a pace of desertification in regions with between 5 and 20 inches of annual rainfall that, if continued, would desertify most of the entire semiarid zone in this century. A decade ago, at the 1996 World Food Summit in Rome, global protection and sustainable management of soil were emphasized as critical for the security of future generations.

Before the Second World War, western Europe was the world's only grain-importing region. Latin American grain exports were nearly double those from North America in the late 1930s. Exports from the Soviet Union's virgin lands were comparable to those from North America's Great Plains. Self-sufficient before the Second World War, Asia, Latin America, eastern Europe, and Africa all now import grain. By the early 1980s over a hundred countries relied on North American grain. Today North America, Australia, and New Zealand are the world's only major grain exporters.

Famine returned to the global scene after decades of unprecedented prosperity in the postwar years when highly variable rainfall, coupled with increasingly severe land degradation, led to regional crop failures. In the mid-1960s, the United States shipped 20 percent of its wheat crop to India to prevent famine from two consecutive crop failures. When Indian crops failed again in 1972, more than eight hundred thousand Indians starved to death. This time there was no American bailout; increased Soviet imports had tied up available wheat supplies. In addition, the 1972 Russian grain purchase encouraged U.S. farmers to plow up marginal land, undermining decades of soil conservation efforts. Today the impact of regional crop failures on global grain prices reflects the close balance between world food supplies and demand. The ongoing availability of surplus North American crops is an issue of global security.

Worldwide, over two billion acres of virgin land have been plowed and brought into agricultural use since 1860. Until the last decades of the twentieth century, clearing new land compensated for loss of agricultural land. In the 1980s the total amount of land under cultivation began declining for the first time since farming reached the land between the Tigris and Euphrates. In the developed world, the rate at which new (and generally marginal) land was brought under cultivation fell below the rate at which land was being exhausted. Although we use a little more than a tenth of Earth's land surface to grow crops, and another quarter of the world's surface for grazing, there is little unused land suitable for either. About the only places left that could be used for agriculture are the tropical forests where thin, highly erodible soils could only briefly support farming.

Because we are already farming about as much of the planet as can be done sustainably, the potential for global warming to affect agricultural systems is alarming. The direct effects of rising temperatures are worrisome enough. A recent study published in the *Proceedings of the National Academy of Sciences* reported that an average daily increase in the growing season's minimum temperature of just 1°C results in a 10 percent reduction in rice yields; similar projections hold for wheat and barley. Beyond the immediate effects on crop yields, global warming scenarios that project anywhere from a 1°C to a 5°C temperature rise over the next century carry a far greater risk.

The world's three great regions of loess soils—the American Midwest, northern Europe, and northern China—produce most of the world's grain. The astounding productivity of modern agriculture depends on the climate of these extensive areas of ideal agricultural soils remaining favorable to crop production. The Canadian and American prairie is already marginal as agricultural land in its western extent. Yet global warming is predicted to increase the severity of droughts here in North America's heartland enough to make that of the Dust Bowl era seem relatively mild. Given the projected doubling of humanity in this century, it is far from certain that the world's population will be able to feed itself.

Other places are predicted to become wetter as global warming leads to a more vigorous hydrological cycle. More frequent high-intensity rainfall events are predicted to substantially increase rainfall erosivity in New England, the mid-Atlantic states, and the Southeast. Models of soil erosion predict from 20 percent to almost 300 percent increases—depending upon how farmers respond to changing rainfall patterns.

Global warming and accelerated erosion are not the only problems facing agricultural land. Growing up in California's Santa Clara Valley, I watched the orchards and fields between Palo Alto and San Jose turn into Silicon Valley. One of the more interesting things I learned from my first job as a foundation inspector was that preparing a building site means carting the topsoil off to a landfill. Sometimes the fine topsoil was sold as fill for use in other projects. Completely paved, Silicon Valley won't feed anyone again for the foreseeable future.

Enough American farms disappeared beneath concrete to cover Nebraska in the three decades from 1945 to 1975. Each year between 1967 and 1977, urbanization converted almost a million acres of U.S. farmland to nonagricultural uses. In the 1970s and 1980s over a hundred acres of U.S. cropland was converted to nonagricultural uses every hour. Urban expansion gobbled up several percent of the best European farmland in the 1960s. Already, urbanization has paved over 15 percent of Britain's agricultural land. The growth of urban areas continues to consume farmland needed to feed cities.

During the cold war, the Department of Agriculture developed tolerance values for soil loss to evaluate the potential for different soils to sustain long-term agricultural production. These values were based on both technical and social inputs—what was considered economically and technically feasible in the 1950s. Soil conservation planning based on this approach typically defines acceptable rates of soil erosion as 5 to 13 tons per hectare per year (2–10 tons per acre per year), equivalent to a loss of an inch of soil in 25 to 125 years (0.2 to 1 mm per year). However, agronomists generally argue that maintaining soil productivity requires keeping erosion under 1 ton per hectare per year—loss of less than one inch in 250 years—2 to 10 times lower than USDA soil loss tolerance values.

Until recent decades, little hard data were available on rates of soil production. So it was hard to know how big a deal to make out of this problem. Farms were losing soil faster than desired, but it was easy to lose sight of the big picture when farmers were struggling to deal with overproduction and food was cheap. Recent studies using a variety of methods, however, all point to soil production rates much lower than USDA soil loss tolerance values. A review of soil production rates from watersheds around the world found rates of from less than 0.1 to 1.9 tons per hectare per year, indicating that the time required to make an inch of soil varies from about 160 years in heather-covered Scotland to more than 4,000 years under deciduous forest in Maryland. Likewise, a global geochemical mass balance based on budgets of the seven major elements in the earth's crust, soils, and waters pegs the average global rate of soil production at between an inch in 240 years to an inch in 820 years (equivalent to an erosion rate of 0.37 to 1.29 tons per hectare per year). For the loess soils of the Great Plains a soil replacement rate of an inch every 500 years is more realistic than the USDA's acceptable soil loss rates. Hence, the currently "acceptable" rates of soil loss are unsustainable over the long run, as they allow soil erosion to proceed four to twenty-five times faster than soil production.

In 1958 the Department of Agriculture found that almost two-thirds of the country's agricultural land was eroding at what it considered a destructive rate—faster than its soil loss tolerance values. A similar survey a decade later found no progress; two-thirds of the country's farmland still lost soil far faster than was acceptable. Despite soil conservation practices promoted after the Dust Bowl, almost two hundred million acres of American farmland were marginalized or lost to crop production by the 1970s. After two centuries of independence, erosion had stripped away a third of the nation's topsoil. At this pace, we would run out of topsoil in less time than has passed since Columbus reached the New World.

By the 1970s many soil conservation plans worked out over the previous decades were abandoned as government policies shifted to support more aggressive cultivation. U.S. farm policy under secretary of agriculture Earl Butz encouraged plowing fence-row-to-fence-row to grow crops to sell to the Russians. Cash crops replaced grass and legumes in crop rotations as bigger tractors increasingly turned soil conservation measures like contour plowing and terracing into annoying nuisances.

In the late 1970s some in Congress viewed with alarm the fact that soil erosion continued to undermine American agriculture despite forty years of effort. The 1977 Soil and Water Resources Conservation Act required the USDA to conduct an intensive appraisal of the nation's soil. Four years in the making, the 1981 report concluded that American soil still eroded at alarming rate more than four decades after the Dust Bowl. In the 1970s the nation lost four billion tons of soil each year—a billion tons a year more than in the 1930s. A freight train filled with all that dirt would stretch around the world twenty four times. At that rate it would take only a century to lose the country's remaining topsoil.

Realistically, political support for spending money to save soil was hard to reconcile with official encouragement to aggressively grow as large a crop as possible to sell overseas. Adjusted for inflation, government support for agricultural conservation programs fell by more than half in the 1970s. No amount of data was going to change congressional perception that the real problem was low prices because of overproduction. Why spend taxpayers' money to save soil when grain bins were bursting?

Part of the problem was that after decades of substantial expenditures on soil conservation programs, there was little solid information on their effectiveness at reducing erosion from America's farms. One of the few well-documented examples of such studies found a substantial reduction in soil erosion from Coon Creek, Wisconsin, for the period from 1936 to 1975.

Designated the nation's first conservation demonstration area in 1933, the Coon Creek basin was severely eroded. Fields plowed in a regular pattern on even steep slopes lacked cover crops, were inadequately manured, and had poor crop rotations. Pastures were overgrazed and eroding. Guided over four decades by the Soil Conservation Service, farmers adopted contour plowing, included cover crops in crop rotations, increased manure applications, and plowed crop residues back into the soil. By 1975 widespread adoption of improved farming practices had reduced hillslope erosion in the basin to just a quarter of what it had been in 1934.

Recent USDA estimates show soil erosion from U.S. cropland as dropping from about three billion tons in 1982 to just under two billion tons in 2001, substantial progress to be sure—but still far ahead of soil production. In the late 1990s Indiana farms still lost a ton of soil to harvest a ton of grain. Even though we know that the soil conservation efforts of ancient civilizations proved too little, too late in case after case, we remain on track to repeat their stories. Only this time we're doing it on a global scale.

Across the planet, moderate to extreme soil erosion has degraded 1.2 billion hectares of agricultural land since 1945—an area the size of China and India combined. One estimate places the amount of agricultural land used and abandoned in the past fifty years as equal to the amount farmed today. The United Nations estimates that 38 percent of global cropland has been seriously degraded since the Second World War. Each year farms around the world lose 75 billion metric tons of soil. A 1995 review of the global effects of soil erosion reported the loss of twelve million hectares of arable land each year to soil erosion and land degradation. This would mean that the annual loss of arable land is almost 1 percent of the total available. Clearly this is not sustainable.

Globally, average cropland erosion of ten to a hundred tons per hectare per year removes soil about ten to a hundred times faster than it forms. So far in the agricultural era, nearly a third of the world's potentially farmable land has been lost to erosion, most of it in the past forty years. In the late 1980s a Dutch-led assessment of global soil erosion found that almost 2 billion hectares of former agricultural lands could no longer support crops. That much land could feed billions of people. We are running out of dirt we cannot afford to lose.

In the mid-1990s David Pimentel's research group at Cornell University estimated the economic costs of soil erosion and the potential economic benefits of soil conservation measures. They considered on-site costs for replacing water-holding capacity lost to soil erosion and for using fertilizers to replace lost soil nutrients. They also estimated off-site costs for increased flood damage, lost reservoir capacity, and dredging of silt-choked rivers to maintain navigation. They estimated that undoing damage caused by soil erosion would cost the United States $44 billion a year, and about $400 billion a year worldwide, more than $70 per person on the planet—higher than the annual income for most people.

Pimentel's group estimated that it would take an annual investment of about $6 billion to bring erosion rates on U.S. cropland into line with soil production. An additional $2 billion a year would do so on U.S. pasturelands. Each dollar invested in soil conservation would save society more than five dollars.

In the short term, though, it can be cheaper for farmers to disregard soil conservation; the cost of reducing soil erosion can be several times the immediate economic benefit of doing so. Farmers with high debt and/or a narrow profit margin can be forced to choose between conserving soil and going bankrupt or working the land until it becomes economically futile. Economic and political incentives encourage practices that destroy soil productivity over the long run, yet preserving the agricultural foundation of civilization requires protecting land from accelerated soil erosion and conversion to other uses.

Many soil conservation measures are proven technologies. Measures adopted to curb soil erosion after the Dust Bowl were not new ideas—contour plowing and cover cropping were known more than a century before. Crop rotations, mulching, and the use of cover crops are ancient ideas. So is terracing, which can reduce erosion by 90 percent, enough to offset the typical increase in erosion rates from cultivation.

Soil conservation trials in Texas, Missouri, and Illinois slowed erosion by a factor of two to a thousand and increased crop yields by up to a quarter for crops like cotton, corn, soybeans, and wheat. Soil conservation is not radical new territory. Many of the most effective methods have been recognized for centuries.

Despite compelling evidence that soil erosion destroyed ancient societies, and can seriously undermine modern societies, some warnings of an impending global soil crisis and food shortages have been overblown. In the early 1980s agricultural economist Lester Brown warned that modern civilization could run out of dirt before oil. Failure of such alarming predictions to play out over the past several decades helped conventional resource economists downplay the potential for soil erosion to compromise food security. Yet such views are shortsighted when erosion removes soil from agricultural fields faster than it forms. Arguing about whether soil loss will become an acute crisis in 2010 or 2100 misses the point.

Analysts offer many reasons for lack of progress in the global war on poverty, but almost every region of acute poverty shares a deteriorating environment. When the productive capacity of the land fails, those living directly off the land suffer most. While land degradation results from economic, social, and political forces, it is also a primary driver of those forces. Increasingly, land degradation is becoming a principal cause of poverty in the developing world. Realistically, the war on poverty simply cannot be won by methods that further degrade the land.

But soil loss is not inevitable. There are productive, profitable farms in every state—and probably every country—that operate with no net loss of topsoil. Despite substantial progress and advances in soil conservation in the past half century, society still prioritizes production over long-term stewardship of the land. The direct costs to farmers from soil erosion in the form of reduced crop yields are typically negligible in the short run, which means conservation measures may never be adopted even if they make economic sense over the long run. We therefore remain in the awkward situation that many highly productive farms mine their own future productivity.

The lessons of the Dust Bowl and the Sahel make a strong case for governments to coordinate, prioritize, and invest in soil conservation. Individuals don't necessarily have an incentive to protect humanity's investment in the soil because their short-term interests need not align with society's long-term interests. A key problem therefore lies in how we view the business of farming. It is the foundation for all other businesses, yet we increasingly treat farming as simply another industrial process.

During the nineteenth century, expansion of the area under cultivation more than kept pace with population growth as pioneer farmers plowed up the Great Plains, the Canadian prairies, the Russian steppe, and vast areas of South America and Australia. Even early in the twentieth century it was clear that further population growth would have to come from increasing crop yields rather than plowing more land.

Together John Deere's plows and Cyrus McCormick's reapers allowed farmers to work far more land than a single farm's livestock could reliably manure. To expand the scope of cultivation and take full advantage of the new equipment required farmers either to continue the pattern of depending on access to fresh land or find a substitute for the eighty head of cattle needed to keep a quarter-section of land well manured. The potential to cultivate far larger areas with new labor-saving machinery provided a ready market for fertilizers. No longer would the scale of an agricultural operation be limited by the capacity of a farm to recycle soil fertility.

EIGHT

Dirty Business

A nation that destroys its soils, destroys itself.

FRANKLIN D. ROOSEVELT

SEVERAL YEARS AFTER SEEING THE RAPID PACE of soil destruction in the lower Amazon I found the antithesis while leading an expedition in eastern Tibet. Driving the region's rough dirt roads I saw a thousand-year-old agricultural system along the valley of the Tsangpo River. We were there to study an ancient ice-dammed lake that drained in a cataclysmic flood down the Himalayan gorge through which the river slices to join the Ganges. Looking for outcrops of ancient lakebeds we drove through villages full of chickens, yaks, and pigs. All around the towns, low silt walls trapped soil in fields of barley, peas, and yellow flowers with seeds rich in canola oil.

After a few days it became obvious that corralling dirt was only part of the secret behind ten centuries of farming the lakebed. Following an unsupervised daily rhythm, Tibetan livestock head out to the fields during the day, fend for themselves, and come home at night. Driving back through towns at the end of each day's fieldwork, we saw pigs and cattle waiting patiently to reenter family compounds. These self-propelled manure dispensers were prolific; even a brief rain turned fields and roads to flowing brown muck.

The night after finding the remains of the glacial dam that once impounded the lake, we stayed at a cheap hotel in the end-of-the-road town of Pai. Homemade sleeping platforms served as beds in sleeping stalls barely separated by unfinished plank walls. The proprietor advised us on our way in that the backyard would serve as our bathroom. That the pigs clean up the yard bothered me during our pork dinner. Still, I had to appreciate the efficiency of pigs eating waste and fertilizing the soil, and then people eating both crops and pigs.

Overlooking the obvious public health issues, this system sustained soil fertility. Other than the occasional satellite dish protruding from the side of a house, villages along the Tsangpo looked much as they had soon after the lake drained. Controlling soil erosion and letting livestock manure the ground allowed generation after generation to plow the same fields.

But Tibetan agriculture is changing. On the road leading out of Lhasa, immigrant Chinese farmers and enterprising Tibetans are setting up irrigated fields and greenhouse complexes. Throughout history, technological innovation has periodically increased agricultural output since the first farmers began turning the earth with sticks before planting. Plows evolved from harnessing animals to pull bigger sticks. Heavy metal plows allowed farmers to cultivate the subsoil once the topsoil eroded away. This not only allowed growing crops on degraded land, it brought more land under cultivation.

Tilling the soil breaks up the ground for planting, helps control weeds, and promotes crop emergence. Although it helps grow desired plants, plowing also leaves the ground bare and unprotected by vegetation that normally absorbs the impact of rainfall and resists erosion. Plowing allows farmers to grow far more food and support more people—at the cost of slowly depleting the supply of fertile dirt.

Agricultural practices evolved as farming methods improved through trial and error. Key innovations included experience with manure and regionally adapted crop rotations. Before mechanized agriculture, farmers cultivated a variety of crops, often by hand on small farms that recycled stubble, manure, and sometimes even human waste to maintain soil fertility. Once farmers learned to rotate peas, lentils, or beans with their primary crops, agricultural settlements could persist beyond the floodplains where nature regularly delivered fresh dirt.

In the Asian tropics, the first few thousand years of rice cultivation involved dryland farming, much as in the early history of wheat. Then about 2,500 years ago, people began growing rice in artificial wetlands, or paddies. The new practice helped prevent the nitrogen-depletion that had plagued tropical farmers because the sluggish water nurtured nitrogen-fixing algae that functioned as living fertilizer. Rice paddies also provided ideal environments for decomposing and recycling human and animal wastes.

A phenomenally successful adaptation, wetland rice cultivation spread across Asia, catalyzing dramatic population growth in regions ill suited for previous farming practices. Yet even though the new system supported more people, most still lived on the brink of starvation. Greater food production didn't mean that the poor had more to eat. It usually meant more people to feed.

Geographer Walter Mallory found no shortage of ideas for addressing China's famines in the early 1920s. Civil engineers proposed controlling rivers to alleviate crop-damaging floods. Agricultural engineers suggested irrigation and land reclamation to increase cultivated acreage. Economists proposed new banking methods to encourage investment of urban capital in rural areas. Others with more overtly political agendas wanted to move people from densely populated regions to the wide-open spaces of Mongolia. Focused on treating symptoms, few addressed the root cause of over-aggressive cultivation of marginal land.

In 1920s China it took almost an acre (0.4 hectares) of land to feed a person for a year. A third of all land holdings were less than half an acre—not enough to feed a single person, let alone support a family. More than half of individual land holdings covered less than an acre and a half, a reality that kept the Chinese at perpetual risk of starvation. A bad year—failure of a single crop—brought famine. China was at the limit of its capacity to feed itself.

Obtaining food consumed 70 to 80 percent of an average family income. Even so, the typical diet consisted of two meals of rice, bread, and salt turnips. People survived from harvest to harvest.

Still, Mallory was impressed that peasant farmers maintained soil fertility despite intensive cultivation for more than four thousand years. He contrasted the longevity of Chinese agriculture with the rapid exhaustion of American soils. The key appeared to be intensive organic fertilization by returning human wastes from cities and towns to the fields. Without access to chemical fertilizers Chinese peasants fertilized the land themselves. By Mallory's time, soil nutrients had been recycled through more than forty generations of farmers and their fields.

In the 1920s famine-relief administrator Y. S. Djang investigated whether people in provinces with abundant harvests ate more food than they needed. It was considered an issue of national concern that some provinces gorged while their neighbors starved.

One remarkable practice Djang found was prevalent in the province of Shao-hsing (Shaoxing), where crops were reliable and abundant. He re-

Figure 23. Chinese farmers plowing sand (courtesy of Lu Tongjing).

ported that people routinely ate more than twice the rice they could digest, stuffing themselves with as many as three "double-strength" portions of rice a day. So the region's human waste made superb fertilizer—and there was lots of it. Even after abundant harvests the population would not sell to outside buyers. Instead, these practical farmers built and maintained elegant public outhouses that served as rice-recapture facilities. They routinely ate surplus crops, reinvesting in their stock of natural capital by returning the partially digested excess to the soil.

Today about a third of China's total cultivated area of 130 million hectares is being seriously eroded by water or wind. Erosion rates in the Loess Plateau almost doubled in the twentieth century; the region now loses an average of more than a billion and a half tons of soil a year. Fully half of the hilly area of the Loess Plateau has lost its topsoil, even though labor-intensive terracing during the Cultural Revolution helped halve the sediment load of the Yellow River.

From the 1950s to the 1970s China lost twenty five million acres of cropland to erosion. Between 20 to 40 percent of southern China's soil has lost its A horizon, reducing soil organic matter, nitrogen, and phosphorus by up to 90 percent. Despite growing use of synthetic fertilizers, Chinese crop yields fell by more than 10 percent from 1999 to 2003. With China starting to run out of farmland, it is unsettling to wonder what might happen were a billion people to start squabbling with their neighbors over food.

On a more optimistic note—as we ponder whether agriculture will be able to keep up with the world's population—we might take comfort in the amazing twentieth-century growth in agricultural production.

Until the widespread adoption of chemical fertilizers, growth in agricultural productivity was relatively gradual. Improvements in equipment, crop rotations, and land drainage doubled both European and Chinese crop yields between the thirteenth and nineteenth centuries. Traditional agricultural practices were abandoned as obsolete when discovery of the elements that form soil nutrients set the stage for the rise of industrial agrochemistry.

Major scientific advances fundamental to soil chemistry occurred in the late eighteenth and early nineteenth centuries. Daniel Rutherford and Antoine Lavoisier respectively discovered nitrogen and phosphorus four years before the American Revolution. Humphrey Davy discovered potassium and calcium in 1808. Twenty years later Friederich Wöhler synthesized urea from ammonia and cyanuric acid, showing it was possible to manufacture organic compounds.

Humphrey Davy endorsed the popular theory that manure helped sustain harvests because organic matter was the source of soil fertility. Then in 1840 Justus von Liebig showed that plants can grow without organic compounds. Even so, Liebig recommended building soil organic matter through manure and cultivation of legumes and grasses. But Liebig also argued that other substances with the same essential constituents could replace animal excrement. "It must be admitted as a principle of agriculture, that those substances which have been removed from a soil must be completely restored to it, and whether this restoration be effected by means of excrements, ashes, or bones, is in a great measure a matter of indifference. A time will come when fields will be manured with a solution . . . prepared in chemical manufactories."[1] This last idea was revolutionary.

Liebig's experiments and theories laid the foundation of modern agrochemistry. He discovered that plant growth was limited by the element in shortest supply relative to the plant's needs. He was convinced that crops could be grown continuously, without fallowing, by adding the right nutrients to the soil. Liebig's discovery opened the door to seeing the soil as a chemical warehouse through which to supply crop growth.

Inspired by Liebig, in 1843 John Bennet Lawes began comparing crop yields from fertilized and unfertilized fields on Rothamsted farm, his family's estate just north of London. An amateur chemist since boyhood, Lawes studied chemistry at Oxford but never finished a degree. Nonetheless, he experimented with agricultural chemistry while running the farm. After investigating the influence of manure and plant nutrients on crop growth, Lawes employed chemist Joseph Henry Gilbert to test whether Liebig's mineral nutrients would keep fields fertile longer than untreated fields. Within a decade it was clear that nitrogen and phosphorus could boost crop yields to match, or even exceed, those from well-manured fields.

An enterprising friend aroused Lawes's curiosity and commercial instincts by asking whether he knew of any profitable use for industrial waste consisting of a mix of animal ashes and bone. Turning waste into gold was the perfect challenge for a frustrated chemist. Natural mineral phosphates are virtually insoluble, and therefore have little immediate value as fertilizer—it takes far too long for the phosphorus to weather out and become usable by plants. But treating rock phosphate with sulfuric acid produced water-soluble phosphates immediately accessible to plants. Lawes patented his technique for making superphosphate fertilizer enriched with nitrogen and potassium and set up a factory on the Thames River in 1843. The dramatic effect of Lawes's product on crop yields meant that by the end of the century Britain was producing a million tons of superphosphate a year.

Bankrolled by substantial profits, Lawes split his time between London and Rothamsted, where he used his estate as a grand experiment to investigate how crops drew nutrition from the air, water, and soil. Lawes oversaw systematic field experiments on the effects of different fertilizers and agricultural practices on crop yields. Not only was nitrogen necessary for plant growth, but liberal additions of inorganic nitrogen-based fertilizer greatly increased harvests. He saw his work as fundamental to understanding the basis for scientific agriculture. His peers agreed, electing Lawes a fellow of the Royal Society in 1854, and awarding him a royal medal in 1867. By the end of the century, Rothamsted was the model for government-sponsored research stations spreading a new agrochemical gospel.

Now a farmer just had to mix the right chemicals into the dirt, add seeds, and stand back to watch the crops grow. Faith in the power of chemicals to catalyze plant growth replaced agricultural husbandry and made both crop rotations and the idea of adapting agricultural methods to the land seem quaint. As the agrochemical revolution overturned practices and traditions developed and refined over thousands of years, large-scale agrochemistry became conventional farming, and traditional practices became alternative farming—even as the scientific basis of agrochemistry helped explain traditional practices.

Nineteenth-century experiments showed that grazing animals process only a quarter to a third of the nitrogen in the plants they ingest. So their dung is full of nitrogen. Still, manure does not return all the nitrogen back to the soil. Without fertilizers, periodically cultivating legumes is the only way to retain soil nitrogen and still harvest crops over the long run. Native cultures around the world independently discovered this basic agricultural truth.

In 1838 Jean-Baptiste Boussingault demonstrated that legumes restored nitrogen to the soil, whereas wheat and oats could not. Here at last was the secret behind crop rotations. It took another fifty years to figure out how it worked. In 1888 a pair of German agricultural scientists, Hermann Hellriegel and Hermann Wilfarth, published a study showing that in contrast to grains, which used up the nitrogen in the soil, legumes were symbiotic with soil microbes that incorporated atmospheric nitrogen into organic matter. By the time the pair of Hermanns figured out the microbial basis for the nitrogen restoring properties of beans, peas, and clover, the agrochemical philosophy was already entrenched, spurred on by the discovery of large deposits of guano off the Peruvian coast.

Peruvians had known of the fertilizing effects of guano for centuries before the conquistadors arrived. When scientific explorer Alexander von Humboldt brought a piece collected from the Chincha Islands back to Europe in 1804 the curious white rock attracted the attention of scientists interested in agricultural chemistry. Situated off the arid coast of Peru, the Chincha Islands provided an ideal environment where huge colonies of nesting seabirds left tons of guano in a climate rainless enough to preserve it. And there was a lot—in places the Chincha guano deposits stood two hundred feet thick, a mountain of stuff better than manure. Phosphate-rich guano also has up to thirty times more nitrogen than most manures.

Recognition of the fertilizing properties of guano led to a nineteenth-century gold rush on small islands composed almost entirely of the stuff. The new system worked well—until the guano ran out. By then the widespread adoption of chemical fertilizers had shifted agricultural practices away from husbandry and nutrient cycling in favor of nutrient application.

The first commercial fertilizer imported to the United States inaugurated a new era in American agriculture when John Skinner, the editor of the *American Farmer,* imported two casks of Peruvian guano to Baltimore in 1824. Within two decades, regular shipments began arriving in New York.

Figure 24. Lithograph of mountainous Chincha Islands guano deposit, circa 1868 (*American Agriculturist* [1868] 27:20).

The guano business boomed. England and the United States together imported a million tons a year by the 1850s. By 1870 more than half a billion dollars' worth of the white gold had been hauled off the Chincha Islands.

As much as conservative agricultural societies scoffed at the notion that bird droppings could revive the soil, farmers who tried it swore by the results. Given the cost and difficulty of obtaining the stuff, the steady spread of guano from Maryland to Virginia and the Carolinas attests to its effect on crop yields. Widespread adoption of guano opened the door for the chemical fertilizers that followed by breaking any dependence on manure to sustain soil fertility. This transformed the basis for farming from a reliance on nutrient recycling into a one-way transfer of nutrients to consumers. From then on nothing came back to the farm.

In the end, only so much guano could be mined from South American islands. Peruvian imports peaked in 1856. By 1870 all the high-quality Chincha guano was gone. In 1881 Bolivia—now the only landlocked country with a navy—lost its Pacific coastline to Chile in a war fought over access to guano islands. Within a few years guano taxes financed the Chilean government. Demonstrated to greatly enhance harvests, guano rapidly became a strategic resource.

The government of Peru maintained tight control over its guano monopoly. American farmers frustrated over the rising price of Chincha Islands guano agitated for breaking the Peruvian monopoly. President Millard Fillmore admonished Congress in 1850 that it was the duty of the government to ensure guano traded at a reasonable price. Entrepreneurs scoured whaling records to rediscover unclaimed guano islands where the stuff could be mined freely. After President Franklin Pierce signed the 1856 Guano Island Act, making it legal for any U.S. citizens to claim any unoccupied guano island as their personal property, several dozen small tropical islands became the United States' first overseas possessions. Paving the way for later global engagements, these diminutive territories helped lead to the development of the modern chemical fertilizer industry.

Industrializing European nations that lacked phosphate deposits raced to grab guano islands. Germany annexed phosphate-rich Nauru in 1888, but lost the island after the First World War when the League of Nations placed it under British administration. In 1901 Britain annexed Ocean Island—a pile of phosphate eight and a half miles square. The British-owned Pacific Islands Company wanted to sell the stuff to Australia and New Zealand, which lacked cheap phosphates. For an annual payment of £50 the company bought the mining rights for the whole island from a local chief with dubious authority. Too lucrative to be inconvenienced by such formalities, the Ocean Island phosphate trade reached one hundred thousand tons a year by 1905.

After the First World War the British Phosphate Commission bought the Pacific Islands Company and increased phosphate mining from Nauru sixfold. In response to the islanders' protests that stripping the island of vegetation and soil was destroying their land, the British government confiscated the remaining lands that could be mined. Shortly thereafter deep mining operations began throughout the island. After that a million tons of phosphate left for commonwealth farms each year. Although Nauru gained independence in 1968, the phosphate deposits are mostly gone and the government is virtually bankrupt. Once a lush paradise, this island nation—the world's smallest republic—has been completely strip-mined. The few remaining islanders live on the coast surrounding the barren moonscape of the island's mined-out interior.

Ocean Island is no better off. Phosphate deposits were exhausted by 1980, leaving the inhabitants to eke out a living from land made uninhabitable to bolster the fertility of foreign soils. The island now specializes as a haven for tax shelters.

Large phosphate deposits were discovered in South Carolina on the eve of the Civil War. Within two decades South Carolina produced more than a third of a million tons of phosphate a year. Southern farmers began combining German potash with phosphoric acid and ammonia to create nitrogen, phosphorous, and potassium based fertilizer to revive cotton belt soils.

The emancipation of slaves spurred the rapid growth in fertilizer's use because plantation owners could not otherwise afford to cultivate their worn-out land with hired labor. Neither could they afford to have large tracts of taxable land lie idle. So most plantation owners rented out land to freed slaves or poor farmers for a share of the crop or a fixed rent. The South's new tenant farmers faced constant pressure to wrest as much as they could from their fields.

Merchants saw tenant farmers trying to work old fields as a captive market for new commercial fertilizers. They were too poor to own livestock, yet their fields would not produce substantial yields without manure. When merchants began lending small farmers the supplies needed to carry them from planting to harvest, experience quickly showed that paying off high-interest, short-term loans required liberal use of commercial fertilizers. Conveniently, bulk fertilizer could be purchased from the merchants who provided the loans in the first place.

Just before the Civil War, Mississippi's new state geologist Eugene Hilgard spent five years touring the state to inventory its natural resources. His 1860 *Report on the Geology and Agriculture of the State of Mississippi* gave birth to modern soil science by proposing that soil was not just leftover dirt made of crumbled rocks but something shaped by its origin, history, and relationship to its environment.

Seeking out virgin soils, Hilgard soon realized that different soils had different characteristic thickness that corresponded to the depth of plant rooting. He described how soil properties changed with depth, defining topsoil and subsoil (what soil scientists now call the A and B horizons) as distinct features. Most radically, Hilgard conceived of soil as a dynamic body transformed and maintained by interacting chemical and biological processes.

Both geologist and chemist by training, Hilgard argued that the secret to fertile soil lay in retaining soil nutrients. "No land can be permanently fertile, unless we restore to it, regularly, the mineral ingredients which our crops have withdrawn."[2] Hilgard admired the Asian practice of returning human waste to the fields to maintain soil fertility by recycling nutrients. He considered America's sewers conduits draining soil fertility to the ocean. Refusing to contribute to this problem, he personally fertilized his own backyard garden.

In an address to the Mississippi Agricultural and Mechanical Fair Association in November 1872, Hilgard spoke of how soil exhaustion shaped the fate of empires. "In an agricultural commonwealth, the fundamental requirement of continued prosperity is . . . that the fertility of the soil must be maintained. . . . The result of the exhaustion of the soil is simply depopulation; the inhabitants seeking in emigration, or in conquest, the means of subsistence and comfort denied them by a sterile soil at home." Hilgard warned that improvident use of the soil would lead America to the same end as Rome.

> Armed with better implements of tillage it takes but a short time to "tire" the soil first taken in cultivation. . . . If we do not use the heritage more rationally, well might the Chickasaws and the Choctaws question the moral right of the act by which their beautiful parklike hunting grounds were turned over to another race, on the plea that they did not put them to the uses for which the Creator intended them. . . . Under their system these lands would have lasted forever; under ours, as heretofore practiced, in less than a century more the State would be reduced to the condition of the Roman Campagna.[3]

Hilgard captivated the audience with his conviction and compelling delivery—until he explained that maintaining soil fertility required applying marl to acidic fields and spreading manure year after year. All that sounded like more trouble than it was worth.

Hilgard rightly dismissed the popular idea that the source of soil fertility lay in the organic compounds in the soil. Also rejecting the western European doctrine that soil fertility was based on soil's texture and its ability to absorb water, he believed that clays retained nutrients necessary for plant growth and considered reliance on chemical fertilizers a dangerous addiction that promoted soil exhaustion.

Hilgard recognized that certain plants revealed the nature of the underlying soil. Crab apple, wild plum, and cottonwoods grew well on calcium-rich soil. Pines grew well on calcium-poor soil. Hired by the federal government to assess cotton production for the 1880 census, he produced two volumes that divided regional soils into distinct classes based on their physical and chemical differences. Hilgard stressed understanding the physical character of a soil, as well as its thickness and the depth to water, before judging its agricultural potential. He thought that phosphorus and potassium in minerals and nitrogen in soil organic matter controlled soil fertility. Hilgard's census report noted that aggressive fertilizer use was starting to revive agriculture in the Carolinas.

He also reported how Mississippi hill country farmers concentrated on plowing valley bottoms where upland dirt had piled up after cotton plantations stripped off the black topsoil. Great gullies surrounded empty manors amidst abandoned upland fields. Hilgard thought that a permanent agriculture required small family farms rather than large commercial plantations or tenant farmers seeking to maximize each year's profits.

With a view of the soil forged in the Deep South, Hilgard moved to Berkeley in his early forties to take a professorship at the new University of California. He arrived just as Californians began shaking off gold rush fever to worry about how to farm the Central Valley's alkali soils—salty ground unlike anything back East. Newspapers were full of accounts of crops that withered mysteriously or produced marginal yields.

The extent of alkali soils increased as irrigation spread across the golden state. Every new irrigated field raised the local groundwater table a little more. Each summer, evaporation pumped more salt up into the soil. Hilgard realized that, like a lamp's wick, clay soils brought the salt closer to the surface. Better drained, sandy soils were less susceptible to salt buildup. Hilgard also realized that alkali soils could make excellent agricultural soils—if you could just get rid of the salt.

Hilgard fought the then popular idea that salty soils resulted from seawater evaporated after Noah's flood. The ancient flood idea simply didn't hold water; the dirt was full of the wrong stuff. California's soils were rich in sodium sulfate and sodium carbonate, whereas seawater was enriched in sodium chloride. The salts in the soil were weathering out of rocks, dissolving in soil water, and then reprecipitating where the water evaporated. He reasoned that drier areas had saltier soil because rain sank into the ground and evaporated in the soil. So just as greater rainfall leached the alkali from the soil, repeated flooding could flush salts from the ground.

Collaborating with farmers eager to improve their land, Hilgard also advocated mulching to reduce evaporation of soil moisture. He experimented with using gypsum to reclaim alkali soils. On New Year's Eve 1893 the *San Francisco Examiner* trumpeted Hilgard's successful transformation of "alkali plains to fields of waving grain." Later that year, on August 13, the *Weekly Colusa Sun* went so far as to assert that Hilgard's work was worth "the whole cost of the University."

Whereas Hilgard's Mississippi work showed the importance of geology, topography, and vegetation to soil development, his California work stressed the importance of climate. In 1892 Hilgard published a landmark report that synthesized data from around the country to explain how soils formed. He explained why soils rich in calcium carbonate typical of the West were unusual in the East, and how greater temperature and moisture in the tropics leached out nutrients to produce thoroughly rotten dirt. Hilgard's report laid out the basic idea that the physical and chemical character of soils reflect the interplay of a region's climate and vegetation working to weather the underlying rocks. Soils were a dynamic interface— literally the skin of the earth.

Before Hilgard's synthesis, soil science was dominated by perceptions based on the humid climates of Europe and the eastern United States. Differences between soils were thought simply to reflect differences in the stuff left over from the dissolution of different rocks. By showing that climate was as important as geology, Hilgard showed that soil was worthy of study in its own right. He also championed the idea that nitrogen was the key limiting nutrient in soils based on observed variations in their carbon to nitrogen ratio and thought that crop production generally would respond greatly to nitrogen fertilization.

Now recognized as one of the founding fathers of soil science, Hilgard's ideas regarding soil formation and nitrogen hunger were ignored in agricultural colleges back East. In particular, South Carolina professor Milton Whitney championed the view that soil moisture and texture alone controlled soil fertility, maintaining that soil chemistry didn't really matter because any soil had more nutrients than required by crops. What was important was the mix of silt, sand, and clay. Based on bulk chemistry, Whitney had a point. But Hilgard knew that not everything in a soil was available to plants.

In 1901 Whitney was appointed chief of the U.S. Department of Agriculture's Bureau of Soils. The new bureau launched a massive national soil and land survey, published detailed soil survey maps for use by farmers, and exuded confidence in the nation's dirt, believing that all soils contained enough inorganic elements to grow any crop. "The soil is the one indestructible, immutable asset that the Nation possesses. It is the one resource that cannot be exhausted; that cannot be used up."[4] Outraged, an aging Hilgard complained about the lack of geologic and chemical information in the new bureau's surveys.

Several years before, in 1903, Whitney had published a USDA bulletin arguing that all soils contained strikingly similar nutrient solutions saturated in relatively insoluble minerals. According to Whitney, soil fertility simply depended on cultural methods used to grow food rather than the native ability of the soil to support plant growth. Soil fertility was virtually limitless. An incensed Hilgard devoted his waning years to battling the politically connected Whitney's growing influence.

A year before he published the controversial bulletin, Whitney had hired Franklin King to head a new Division of Soil Management. A graduate of Cornell University, King had been appointed in 1888 by the University of Wisconsin to be the country's first professor of agricultural physics at the age of forty. Considered the father of soil physics in the United States, King had also studied soil fertility.

King's stay in Washington was short. In his new post, King studied relations between bulk soil composition, the levels of plant nutrients in soil solutions, and crop yields. He found that the amount of nutrients in soil solutions differed from amounts suggested by total chemical analysis of soil samples but correlated with crop yields—conclusions at odds with those published by his new boss. Refusing to endorse King's results, Whitney forced him to resign from the bureau and return to academia where he would be less of a nuisance.

While Hilgard and Whitney feuded in academic journals, a new concept evolved of soils as ecological systems influenced by geology, chemistry, meteorology, and biology. In particular, recognition of the biological basis for nitrogen fixation helped lay the foundation for the modern concept of the soil as the frontier between geology and biology. Within a century of their discovery, nitrogen, phosphorus, and potassium were recognized to be the key elements of concern to agriculturalists. How to get enough of them was the issue.

Even though nitrogen makes up most of our atmosphere, plants can't use nitrogen bound up as stable N_2 gas. In order to be used by organisms, the inert double nitrogen molecule must first be broken and the halves combined with oxygen, carbon, or hydrogen. The only living organisms capable of doing this are about a hundred genera of bacteria, those associated with the roots of legumes being the most important. Although most crops deplete the supply of nitrogen in the soil, root nodules on clover, alfalfa, peas, and beans house bacteria that make organic compounds from atmospheric nitrogen. This process is as essential to us as it is to plants because we need to eat ten preformed amino acids we can't assemble ourselves. Maintaining high nitrogen levels in agricultural soil requires rotating crops that consume nitrogen with crops that replenish nitrogen—or continually adding nitrogen fertilizers.

Phosphorus is not nearly as abundant as nitrogen, but it too is essential for plant growth. Unlike potassium, which accounts for an average of 2.5 percent of the earth's crust and occurs in rocks almost everywhere in forms readily used as natural fertilizer, phosphorus is a minor constituent of rock-forming minerals. In many soils, its inaccessibility limits plant growth. Consequently, phosphorus-based fertilizers greatly enhance a crop's productivity. The only natural sources of phosphorus other than rock weathering are relatively rare deposits of guano or more common but less concentrated calcium-phosphate rock. By 1908 the United States was the largest single producer of phosphate in the world, mining more than two and a half million tons from deposits in South Carolina, Florida, and Tennessee. Almost half of U.S. phosphate production was exported, most of it to Europe.

By the First World War serious depletion of phosphorus was apparent in American soils.

> For extensive areas in the South and East the phosphorus is so deficient that there is scarcely any attempt to raise a crop without the use of phosphate compounds as fertilizers. . . . Western New York and Ohio, which not more than fifty or sixty years ago were regarded as the very center of the fertility of the country, are very seriously depleted in this element; and into them there is continuous importation of phosphate fertilizer.[5]

Early twentieth-century estimates of the amount of phosphorus lost in typical agricultural settings predicted that a century of continuous cropping would exhaust the natural supply in midwestern soils. As phosphate became a strategic resource, calls for nationalizing phosphate deposits and prohibiting exports began to circulate in Washington.

On March 12, 1901, the United States Industrial Commission invited Bureau of Soils chief Milton Whitney to testify about abandoned farmland in New England and the South. Whitney attributed New England's abandoned farms to the falling price of crops pouring out of the Midwest on the nation's new railroads. In his opinion, New England's farmers just could not compete with cheap wheat and cattle from out West.

Whitney told the committee that growing crops poorly suited to a region's soil or climate led to abandoned farms. He described how farms established twenty years earlier in semiarid parts of Kansas, Nebraska, and Colorado had experienced boom times for a few years, and then failed after a run of dry years. Whitney was certain that it would happen again given the region's unpredictable rainfall.

Whitney also thought that social conditions affected farm productivity. Prime farmland in southern Maryland sold for about ten dollars an acre. Similar land in Lancaster County, Pennsylvania, sold for more than ten times as much. Since Whitney believed that all soils were capable of similar productivity, he invoked social factors to explain differences in land values. Pennsylvanian farmers owned their farms and grew a diverse array of crops, including most of their own food. They sold their surplus locally. In contrast, hired overseers or tenant farmers worked Maryland's farms growing tobacco, wheat, and corn for distant markets. Whitney considered export-oriented, cash-crop monoculture responsible for impoverishing Maryland, Virginia, and the southern states in general.

Whitney saw that fertilizers could greatly increase crop yields. He considered natural fertility to be sustained by rock weathering that produced soil. Fertilizers added extra productivity. "We can unquestionably force the fertility far beyond the natural limit and far beyond the ordinary limits of crop production. . . . In this sense the effect of fertilization is a simple addition of plant food to the soil in such form that the crops can immediately use it."[6] Whitney thought fertilizers sped the breakdown of soil minerals, accelerating soil production. Pumped up on fertilizers, the whole system could run faster.

In effect, Whitney conceived of the soil as a machine that required tuning in order to sustain high crop yields. He thought that American farmers' destructive habit of ignoring the particular type of soil in their fields reflected the fact that they didn't stay on their land very long. In 1910 more than half of America's farmers had been on their land for less than five years, not long enough to get to know their dirt.

Here was where soil scientists could help. "The soil scientist has the same relation to the partnership between the man and the soil that . . . the chemist has to the steel or dye manufacturer." Whitney literally considered soil a crop factory. "Each soil type is a distinct, organized entity—a factory, a machine—in which the parts must be kept fairly adjusted to do efficient work."[7] However, he was unimpressed with how American farmers ran the nation's dirt factories. In Whitney's view, new technologies and more intensive agrochemistry would define America's future. The Bureau of Soils chief did not realize that it would be a British idea implemented with German technology.

In 1898 the president of the British Association, Sir William Crookes, addressed the association's annual meeting, choosing to focus on what he called the wheat problem—how to feed the world. Crookes foresaw the need to radically restructure fertilizer production because society could not indefinitely mine guano and phosphate deposits. He realized that higher wheat yields would require greater fertilizer inputs and that nitrogen was the key limiting nutrient. The obvious long-term solution would be to use the virtually unlimited supply of nitrogen in the atmosphere. Feeding the growing world population in the new century would require finding a way to efficiently transform atmospheric nitrogen into a form plants could use. Crookes believed that science would figure out how to bypass legumes. "England and all civilised nations stand in deadly peril of not having enough to eat. . . . Our wheat-producing soil is totally unequal to the strain put upon it. . . . It is the chemist who must come to rescue. . . . It is through the laboratory that starvation may ultimately be turned into plenty."[8] Ironically, solving the nitrogen problem did not eliminate world hunger. Instead the human population swelled to the point where there are more hungry people alive today than ever before.

In addition to being natural fertilizers, nitrates are essential for making explosives. By the early twentieth century, industrial nations were becoming increasingly dependent on nitrates to feed their people and weapons. Britain and Germany in particular were aggressively seeking reliable sources of nitrates. Both countries had little additional cultivatable land and already imported large amounts of grain, despite relatively high crop yields from their own fields.

Vulnerable to naval blockades that could disrupt nitrate supplies, Germany devoted substantial effort toward developing new methods to capture atmospheric nitrogen. On July 2, 1909, after years of attempting to synthesize ammonia, Fritz Haber succeeded in sustaining production of liquid ammonia for five hours in his Karlsruhe laboratory. Crookes's challenge had been met in just over a decade. Less than a century later, half the nitrogen in the world's people comes from the process that Haber pioneered.

Badische Anilin- und Sodafabrik (BASF) chemist Carl Bosch commercialized Haber's experimental process, now known as the Haber-Bosch process, with amazing rapidity. A prototype plant was operating a year later, construction of the first commercial plant began in 1912, and the first commercial ammonia flowed in September the following year. By the start of the First World War, the plant was capturing twenty metric tons of atmospheric nitrogen a day.

As feared by the German high command, the British naval blockade cut off Germany's supply of Chilean nitrates in the opening days of the war. It soon became clear that the unprecedented amounts of explosives used in the new style of trench warfare would exhaust German munitions in less than a year. The blockade also cut off BASF from its primary markets and revenue sources. Within months of the outbreak of hostilities the company's new ammonia plant was converted from producing fertilizer to nitrates for Germany's ammunition factories. By the war's end, all of BASF's production was used for munitions and together with the German war ministry the company was building a major plant deep inside Germany, safe from French air raids. In the end, however, the German military did not so much run out of ammunition as it ran out of food.

After the war, other countries adopted Germany's remarkable new way of producing nitrates. The Allies immediately recognized the strategic value of the Haber-Bosch process; the Treaty of Versailles compelled BASF to license an ammonia plant in France. In the United States, the National Defense Act provided for damming the Tennessee River at Mussel Shoals to generate cheap electricity for synthetic nitrogen plants that could manufacture either fertilizers or munitions, depending on which was in greater demand.

In the 1920s German chemists modified the Haber-Bosch process to use methane as the feedstock for producing ammonia. Because Germany lacked natural gas fields, the more efficient process was not commercialized until 1929 when Shell Chemical Company opened a plant at Pittsburg, California to convert cheap natural gas into cheap fertilizer. The technology for making ammonia synthesis the dominant means of fixing atmospheric nitrogen arrived just in time for the industrial stagnation of the Depression.

Ammonia plant construction began again in earnest in the run-up to the Second World War. The Tennessee Valley Authority's (TVA) dams provided ideal sites for additional ammonia plants built to manufacture explosives. One plant was operating when Japan bombed Pearl Harbor; ten were operating by the time Berlin fell.

After the war, governments around the world sought and fostered markets for ammonia from suddenly obsolete munitions factories. Fertilizer use in the TVA region shot up rapidly thanks to abundant supplies of cheap nitrates. American fertilizer production exploded in the 1950s when new natural gas feedstock plants in Texas, Louisiana, and Oklahoma were connected to pipelines to carry liquid ammonia north to the corn belt. Europe's bombed-out plants were rebuilt and converted to fertilizer production. Expansion of Russian ammonia production was based on central Asian and Siberian natural gas fields. Global production of ammonia more than doubled in the 1960s and doubled again in the 1970s. By 1998 the world's chemical industry produced more than 150 million metric tons of ammonia a year; the Haber-Bosch process supplied more than 99 percent of production. Natural gas remains the principal feedstock for about 80 percent of global ammonia production.

The agricultural output of industrialized countries roughly doubled in the second half of the twentieth century. Much of this newfound productivity came from increasing reliance on manufactured fertilizers. Global use of nitrogen fertilizers tripled between the Second World War and 1960, tripled again by 1970, and then doubled once more by 1980. The ready availability of cheap nitrogen led farmers to abandon traditional crop rotations and periodic fallowing in favor of continuous cultivation of row crops. For the period from 1961 to 2000, there is an almost perfect correlation between global fertilizer use and global grain production.

Soil productivity became divorced from the condition of the land as industrialized agrochemistry ramped up crop yields. The shift to large-scale monoculture and increasing reliance on fertilizer segregated animal husbandry from growing crops. Armed with fertilizers, manure was no longer needed to maintain soil fertility.

Much of the increased demand for nitrogen fertilizer reflects the adoption of new high-yield strains of wheat and rice developed to feed the world's growing population. In his 1970 Nobel Peace Prize acceptance speech, Norman Borlaug, pioneering developer of the green revolution's high-yield rice, credited synthetic fertilizer production for the dramatic increases in crop production. "If the high-yielding dwarf wheat and rice varieties are the catalysts that have ignited the Green Revolution, then chemical fertilizer is the fuel that has powered its forward thrust."[9] In 1950 high-income countries in the developed world accounted for more than 90 percent of nitrogen fertilizer consumption; by the end of the century, low-income developing countries accounted for 66 percent.

In developing nations, colonial appropriation of the best land for export crops meant that increasingly intensive cultivation of marginal land was necessary to feed growing populations. New high-yield crop varieties increased wheat and rice yields dramatically in the 1960s, but the greater yields required more intensive use of fertilizers and pesticides. Between 1961 and 1984 fertilizer use increased more than tenfold in developing countries. Well-to-do farmers prospered while many peasants could not afford to join the revolution.

The green revolution simultaneously created a lucrative global market for the chemicals on which modern agriculture depended and practically ensured that a country embarked on this path of dependency could not realistically change course. In individuals, psychologists call such behavior addiction.

Nonetheless, green revolution crops now account for more than three-quarters of the rice grown in Asia. Almost half of third-world farmers use green revolution seeds, which doubled the yield per unit of nitrogen fertilizer. In combination with an expansion of the area under cultivation, the green revolution increased third-world agricultural output by more than a third by the mid-1970s. Once again, increased agricultural yields did not end hunger because population growth kept pace—this time growing well beyond what could be maintained by the natural fertility of the soil.

Between 1950 and the early 1970s global grain production nearly doubled, yet per capita cereal production increased by just a third. Gains slowed after the 1970s when per capita grain production fell by more than 10 percent in Africa. By the early 1980s population growth consumed grain surpluses from expanded agricultural production. In 1980 world grain reserves dropped to a forty-day supply. With less than a year's supply of grain on hand, the world still lives harvest to harvest. In developed nations, modern food distribution networks typically have little more than a few days' supply in the pipeline at any one time.

From 1970 to 1990 the total number of hungry people fell by 16 percent, a decrease typically credited to the green revolution. However, the largest drop occurred in communist China, beyond the reach of the green revolution. The number of hungry Chinese fell by more than 50 percent, from more than 400 million to under 200 million. Excluding China, the number of hungry people increased by more than 10 percent. The effectiveness of the land redistribution of the Chinese Revolution at reducing hunger shows the importance of economic and cultural factors in fighting hunger. However we view Malthusian ideas, population growth remains critical—outside of China, increased population more than compensated for the tremendous growth in agricultural production during the green revolution.

Another key reason why the green revolution did not end world hunger is that increased crop yields depended on intensive fertilizer applications that the poorest farmers could not afford. Higher yields can be more profitable to farmers who can afford the new methods, but only if crop prices cover increased costs for fertilizers, pesticides, and machinery. In third world countries the price of outlays for fertilizers and pesticides increased faster than green revolution crop yields. If the poor can't afford to buy food, increased harvests won't feed them.

More ominously, the green revolution's new seeds increased third-world dependence on fertilizers and petroleum. In India agricultural output per ton of fertilizer fell by two-thirds while fertilizer use increased sixfold. In West Java a two-thirds jump in outlays for fertilizer and pesticides swallowed up profits from the resulting one-quarter increase in crop yields in the 1980s. Across Asia fertilizer use grew three to forty times faster than rice yields. Since the 1980s falling Asian crop yields are thought to reflect soil degradation from increasingly intensive irrigation and fertilizer use.

Without cheap fertilizers—and the cheap oil used to make them—this productivity can't be sustained. As oil prices continue climbing this century, this cycle may stall with disastrous consequences. We burned more than a trillion barrels of oil over the past two decades. That's eighty million barrels a day—enough to stack to the moon and back two thousand times. Making oil requires a specific series of geologic accidents over inconceivable amounts of time. First, organic-rich sediment needs to be buried faster than it can decay. Then the stuff needs to get pushed miles down into the earth's crust to be cooked slowly. Buried too deep or cooked too fast and the organic molecules burn off; trapped too shallow or not for long enough and the muck never turns into oil. Finally, an impermeable layer needs to seal the oil in a porous layer of rock from which it can be recovered. Then somebody has to find it and get it out of the ground. It takes millions of years to produce a barrel of oil; we use millions of barrels a day. There is no question that we will run out of oil—the only question is when.

Estimates for when petroleum production will peak range from before 2020 to about 2040. Since such estimates do not include political or environmental constraints, some experts believe that the peak in world oil production is already at hand. Indeed, world demand just rose above world supply for the first time. Exactly when we run out will depend on the political evolution of the Middle East, but regardless of the details oil production is projected to drop to less than 10 percent of current production by the end of the century. At present, agriculture consumes 30 percent of our oil use. As supplies dwindle, oil and natural gas will become too valuable to use for fertilizer production. Petroleum-based industrial agriculture will end sometime later this century.

Not surprisingly, agribusiness portrays pesticide and fertilizer intensive agriculture as necessary to feed the world's poor. Even though almost a billion people go hungry each day, industrial agriculture may not be the answer. Over the past five thousand years population kept pace with the ability to feed people. Simply increasing food production has not worked so far, and it won't if population growth keeps up. The UN Food and Agriculture Organization reports that farmers already grow enough to provide 3,500 calories a day to every person on the planet. Per capita food production since the 1960s has increased faster than the world's population. World hunger persists because of unequal access to food, a social problem of distribution and economics rather than inadequate agricultural capacity.

One reason for the extent of world hunger is that industrialized agriculture displaced rural farmers, forcing them to join the urban poor who cannot afford an adequate diet. In many countries, much of the traditional farmland was converted from subsistence farms to plantations growing high-value export crops. Without access to land to grow their own food, the urban poor all too often lack the money to buy enough food even if it is available.

The USDA estimates that about half the fertilizer used each year in the United States simply replaces soil nutrients lost by topsoil erosion. This puts us in the odd position of consuming fossil fuels—geologically one of the rarest and most useful resources ever discovered—to provide a substitute for dirt—the cheapest and most widely available agricultural input imaginable.

Traditional rotations of grass, clover, or alfalfa were used to replace soil organic matter lost to continuous cultivation. In temperate regions, half the soil organic matter commonly disappears after a few decades of plowing. In tropical soils, such losses can occur in under a decade. By contrast, experiments at Rothamsted from 1843 to 1975 showed that plots treated with farmyard manure for more than a hundred years nearly tripled in soil nitrogen content whereas nearly all the nitrogen added in chemical fertilizers was lost from the soil—either exported in crops or dissolved in runoff.

More recently, a fifteen-year study of the productivity of maize and soybean agriculture conducted at the Rodale Institute in Kutztown, Pennsylvania showed no significant differences in crop yields where legumes or manure were used instead of synthetic fertilizers and pesticides. The soil carbon content for manured plots and those with a legume rotation respectively increased to three to five times that of conventional plots. Organic and conventional cropping systems produced similar profits, but industrial farming depleted soil fertility. The ancient practice of including legumes in crop rotations helped retain soil fertility. Manuring actually increased soil fertility.

This is really not so mysterious. Most gardeners know that healthy soil means healthy plants that, in turn, help maintain healthy soil. I've watched this process in our own yard as my wife doused our lot with soil soup brewed in our garage and secondhand coffee grounds liberated from behind our neighborhood coffee shop. I marvel at how we are using organic material imported from the tropics, where there are too few nutrients in the soil in the first place, to help rebuild the soil on a lot that once had a thick forest soil. Now, five years into this experiment, the soil in our yard has a surface layer of rich organic matter, stays moist long after it rains, and is full of coffee-colored worms.

Our caffeinated worms have been busy since we hired a guy with a small bulldozer to rip out the ragged, eighty-two-year-old turf lawn our house came with and reseed the yard with a mix of four different kinds of plants, two grasses and two forbs—one with little white flowers and the other with little red flowers. The flowers are a nice upgrade from our old lawn and we don't have to water it. Better still, the combination of four plants that grow and bloom at different times keeps out weeds.

Our eco-lawn may be advertised as low maintenance, but we still have to mow it. So we just cut the grass and leave it to rot where it falls. Within a week all the cuttings are gone—dragged down into worm burrows. Now I can dig a hole in the lawn and find big fat worms where there used to be nothing but dry dirt. After just a few years, the ground around the edges of the lawn stands a quarter of an inch higher than the patio surface built at the same time we seeded the eco-lawn. The worms are pumping up the yard—plowing it, churning it, and pushing carbon down into the ground—turning our dirt into soil. Recycling organic matter literally put life back in our yard. Adjusted for scale, the same principles could work for farms.

About the same time that mechanization transformed conventional agriculture, the modern organic farming movement began to coalesce around the ideas of Sir Albert Howard and Edward Faulkner. These two gentlemen with very different backgrounds came to the same conclusion: retaining soil organic matter was the key to sustaining high intensity farming. Howard developed a method to compost at the scale of large agricultural plantations, whereas Faulkner devised methods to plant without plowing to preserve a surface layer of organic matter.

At the close of the 1930s Howard began to preach the benefits of maintaining soil organic matter as crucial for sustaining agricultural productivity. He feared that increasing reliance on mineral fertilizers was replacing soil husbandry and destroying soil health. Based on decades of experience on plantations in India, Howard advocated incorporating large-scale composting into industrial agriculture to restore and maintain soil fertility.

In Howard's view, farming should emulate nature, the supreme farmer. Natural systems provide a blueprint for preserving the soil—the first condition of any permanent system of agriculture. "Mother earth never attempts to farm without live stock; she always raises mixed crops; great pains are taken to preserve the soil and to prevent erosion; the mixed vegetable and animal wastes are converted into humus; there is no waste; the processes of growth and the processes of decay balance one another."[10] Constant cycling of organic matter through the soil coupled with weathering of the subsoil could sustain soil fertility. Preservation of humus was the key to sustaining agriculture.

Howard felt that soil was an ecological system in which microbes provided a living bridge between soil humus and living plants. Maintaining humus was essential for breaking down organic and mineral matter needed to feed plants; soil-dwelling microorganisms that decompose organic matter lack chlorophyll and draw their energy from soil humus. Soil organic matter was essential for the back half of the cycle of life in which the breakdown of expired life fueled the growth of new life.

In the 1920s at the Institute of Plant Industry in Indore, India, Howard developed a system to incorporate composting into plantation agriculture. His process mixed plant and animal wastes to favor the growth of microorganisms, which he considered tiny livestock that enriched the soil by breaking organic matter into its constituent elements. Field trials of

Howard's methods in the tropics were extremely successful. As word of his increased crop yields and soil-building methods spread, plantations in India, Africa, and Central America began adopting his approach.

Howard saw intensive organic farming as how to undo the damage industrial farming inflicted on the world's soils. He thought that many plant and animal diseases arose from reliance on artificial fertilizers that disrupted the complex biology of native soils. Reestablishing organic-rich topsoil through intensive composting would reduce, if not eliminate, the need for pesticides and fertilizers while increasing the health and resilience of crops.

After the First World War, Howard saw munitions factories begin manufacturing cheap fertilizers advertised as containing everything various crops needed. He worried that adopting fertilizers as standard practice on factory farms would emphasize maximizing profits at the expense of soil health. "The restoration and maintenance of soil fertility has become a universal problem. . . . The slow poisoning of the life of the soil by artificial manures is one of the greatest calamities which has befallen agriculture and mankind."[11] The Second World War derailed adoption of Howard's ideas. After the war the companies that supplied the world's armies turned to pumping out fertilizer, this time cheap enough to eclipse soil husbandry.

In the middle of the Second World War, Edward Faulkner published *Plowman's Folly* in which he argued that plowing—long considered the most basic act of farming—was counterproductive. Enrolled in courses on soil management and farm machinery decades earlier at the University of Kentucky, Faulkner had annoyed his professors by asking what was the point of tearing apart the soil for planting instead of incorporating crops into the organic layer at the ground surface where plants naturally germinate. Despite the usual reasons offered for plowing—preparing the seedbed, incorporating crop residues and manure or fertilizers into the soil, and allowing the soil to dry out and warm up in spring—his instructors sheepishly admitted that they knew of no clear scientific reasons for why the first step in the agricultural process was actually necessary. After twenty-five years working as a county agricultural agent in Kentucky and Ohio, Faulkner eventually concluded that plowing created more problems than it solved.

Challenging agronomists to reconsider the necessity of plowing, he argued that the key to growing abundant crops was maintaining an adequate surface layer of organic material to prevent erosive runoff and maintain soil nutrients. This was heresy. "We have equipped our farmers with a greater tonnage of machinery per man than any other nation. Our agricultural population has proceeded to use that machinery to the end of destroying the soil in less time than any other people has been known to do in recorded history."[12] Faulkner also considered reliance on mineral fertilizers unnecessary and unsustainable.

Like most heretics', Faulkner's unconventional beliefs were grounded in experience. He inadvertently discovered in his backyard garden that he could greatly increase crop yields by not tilling when he began growing corn in soil he considered better suited for making bricks. From 1930 to 1937 he introduced organic matter into his backyard plot by digging a trench with a shovel and mixing in leaves at the bottom of the trench to emulate the standard practice of plowing under last year's crop stubble. Like conventional plowing, this buried the organic-rich surface material to a depth of six or eight inches. In the fall of 1937 he tried something different. He mixed the leaves into the surface of the soil.

The next year his soil was transformed. Previously he had been able to grow only parsnips in the stiff clay soil; now the soil texture was granular. It could be raked like sand. In addition to parsnips, he harvested fine crops of carrots, lettuce, and peas—without fertilizer and with minimal watering. All he did was keep the weeds down.

When the Soil Conservation Service staff were unimpressed with his backyard experiment, Faulkner took up the challenge and leased a field for a full-scale demonstration. Instead of plowing before planting, he disked the standing plants into the surface of the soil, leaving the ground littered with chopped-up weeds. Skeptical neighbors forecast a poor harvest for the careless amateur. Surprised and impressed when Faulkner's crop exceeded their own, they were unsure what to think about his mysterious success without plowing, fertilizers, or pesticides.

After several years of repeated success on his leased field, Faulkner began to advocate rebuilding surface layers of organic material. He was confident that with the right approach and machinery, farmers could recreate good soil wherever it had existed naturally. "Men have come to feel . . . that centuries are necessary for the development of a productive soil. The satisfying truth is that a man with a team or a tractor and a good disk harrow can mix into the soil, in a matter of hours, sufficient organic material to accomplish results equal to what is accomplished by nature in decades." What farmers needed to do was stop tilling the soil and begin incorporating organic matter back into the ground. "Everywhere about us is evidence that the undisturbed surface of the earth produces a healthier growth than that portion now being farmed. . . . The net effect of fertilizing the land, then, is not to increase the possible crop yield, but to decrease the devastating effects of plowing."[13] Like Howard, Faulkner believed that reestablishing healthy soil would reduce, if not eliminate, crop pests and diseases.

Soil organic matter is essential for sustaining soil fertility not so much as a direct source of nutrients but by supporting soil ecosystems that help promote the release and uptake of nutrients. Organic matter helps retain moisture, improves soil structure, helps liberate nutrients from clays, and is itself a source of plant nutrients. Loss of soil organic matter reduces crop yields by lowering the activity of soil biota, thereby slowing nutrient recycling.

Different soils in different climates can sustain agriculture without supplemental fertilization for different periods of time. Organic-rich soil of the Canadian Great Plains can be cultivated for more than fifty years before losing half its soil carbon, whereas Amazon rainforest soils can lose all agricultural potential in under five years. A twenty-four-year fertilization experiment in northwestern China found that soil fertility declined under chemical fertilizers unless coupled to addition of straw and manure.

Nowhere is the debate over the appropriate application of technology more polarized than in the field of biotechnology. Downplaying notions of population control and land reform, industry advocates push the idea that genetic engineering will solve world hunger. Despite altruistic rhetoric, genetic engineering companies design sterile crops to ensure that farmers—large agribusinesses and subsistence farmers alike—must keep on buying their proprietary seeds. There was a time when prudent farmers kept their best seed stock for next year's crop. Now they get sued for doing so.

Despite the dramatically increased yields promised by industry, a study by the former director of the National Academy of Sciences' Board on Agriculture found that genetically modified soybean seeds produced smaller harvests than natural seeds when he analyzed more than eight thousand field trials. A USDA study found no overall reduction in pesticide use associated with genetically engineered crops, even though increased pest resistance is touted as a major advantage of crop engineering. Whereas the promise of greatly increased crop yields from genetic engineering has proven elusive, some fear that genetically modified genes that convey sterility could cross to nonproprietary crops, with catastrophic results.

Given the significant real and potential drawbacks of bioengineering and agrochemistry, alternative approaches deserve a closer look. Over the long run, intensive organic farming and other nonconventional methods may prove our best hope for maintaining food production in the face of population growth and continuing loss of agricultural land. In principle, intensive organic methods could even replace fertilizer-intensive agriculture once cheap fossil fuels are history.

Here is the crux of Wes Jackson's argument that tilling the soil has been an ecological catastrophe. A genetics professor before he resigned to become president of the Land Institute in Salina, Kansas, Jackson says he is not advocating a return to the bow and arrow. He just questions the view that plowing the soil is irrefutably wholesome, pointedly suggesting that the plow destroyed more options for future generations than did the sword and that—with rare exceptions—plow-based agriculture hasn't proven sustainable. He estimates that in the next two decades severe soil erosion will destroy 20 percent of the natural agricultural potential of our planet to grow crops without fertilizer or irrigation.

Yet Jackson is neither doomsayer nor Luddite. In person he sounds more like a farmer than an environmental extremist. Instead of despair, he calls for agricultural methods based on emulating natural systems rather than controlling, or replacing them. In promoting natural systems agriculture, Jackson is the latest prophet for Xenophon's philosophy of adapting agriculture to the land rather than vice versa.

Drawing on experience in the American farm belt, Jackson seeks to develop an agricultural system based on imitating native prairie ecosystems. Unlike annual crops raised on bare, plowed ground, the roots of native perennials hold the soil together through drenching rainstorms. Native prairies contain both warm-season and cold-season grasses, as well as legumes and composites. Some of the plants do better in wet years, some thrive in dry years. The combination helps keep out weeds and invasive species because plants cover the ground all year—just like our eco-lawn.

As ecologists know, diversity conveys resilience—and resilience, Jackson says, can help keep agriculture sustainable. Hence he advocates growing a combination of crops year-round to shield the ground from the rain's erosive impact. Monocultures generally leave the ground bare in the spring, exposing vulnerable soil to erosion for months before crops get big enough to block incoming rain. Storms hitting before crops leaf out cause two to ten times the erosion of storms later in the year when the ground lies shielded beneath crops. Under monoculture, one good storm at the wrong time can send erosion racing decades ahead of soil production.

The beneficial effects of Jackson's system are evident at the Land Institute. Research there has shown that a perennial polyculture can manage pests, provide all its own nitrogen, and produce a greater per-acre crop yield than monocultures. Although the specifics of Jackson's approach were designed for the prairies, his methods could be adapted to other regions by using species mixtures appropriate to the local environment. Understandably, pesticide, fertilizer, and biotechnology companies are not very excited about Jackson's low-tech approach. But Jackson is not a lone voice in the wilderness; over the past few decades many farmers have adopted methods like those advocated by Faulkner and Howard.

Whatever we call it, today's organic farming combines conservation-minded methods with technology but does not use synthetic pesticides and fertilizers. Instead, organic farming relies on enhancing and building soil fertility by growing diversified crops, adding animal manure and green compost, and using natural pest control and crop rotation. Still, for a farm to survive in a market economy it must be profitable.

Long-term studies show that organic farming increases both energy efficiency and economic returns. Increasingly, the question appears not to be whether we can afford to go organic. Over the long run, we simply can't afford not to, despite what agribusiness interests will argue. We can greatly improve conventional farming practices from both environmental and economic perspectives by adopting elements of organic technologies. Oddly, our government subsidizes conventional farming practices, whereas the market places a premium on organic produce. A number of recent studies report that organic farming methods not only retain soil fertility in the long term, but can prove cost effective in the short term.

In 1974, under the leadership of ecologist Barry Commoner, the Center for the Biology of Natural Systems at Washington University in St. Louis began comparing the performance of organic and conventional farms in the Midwest. Pairing fourteen organic farms with conventional farms of similar size, operated with a similar crop-livestock system on similar soils, the two-year study found that organic farms produced about the same income per acre as did conventional farms. Although the study's preliminary results surprised skeptical agricultural experts, many subsequent studies confirmed that substantially lower production costs more than offset slightly smaller harvests from organic farms. Industrial agrochemistry is a societal convention and not an economic imperative.

Subsequent studies also showed that crop yields are not substantially lower under organic agricultural systems. Just as important is the demonstration that modern agriculture need not deplete the soil. Rothamsted, the estate where John Lawes proved the fertilizing effects of chemical fertilizers, hosts the longest ongoing comparison of organic and conventional agriculture—a century and a half—placing manure-based organic farming and chemical fertilizer-based farming side by side. Wheat yields from conventionally fertilized and organic plots were within 2 percent of each other, but the soil quality measured in terms of carbon and nitrogen levels improved over time in the organic plots.

A twenty-two-year study by the Rodale Institute on a Pennsylvania farm compared the inputs and production from conventional and organic plots. Average crop yields were comparable under normal rainfall, but the average corn yields in the organic plots were about a third higher during the driest five years. Energy inputs were about a third lower, and labor costs about a third higher in the organic plots. Overall, organic plots were more profitable than the conventional plots because total costs were about 15 percent lower, and organic produce sold at a premium. Over the two-decade-long experiment, soil carbon and nitrogen contents increased in the organic plots.

In the mid-1980s, researchers led by Washington State University's John Reganold compared the state of the soil, erosion rates, and wheat yields from two farms near Spokane in eastern Washington. One farm had been managed without the use of commercial fertilizers since first plowed in 1909. The adjoining farm was first plowed in 1908 and commercial fertilizer was regularly applied after 1948.

Surprisingly, there was little difference in the net harvest between the farms. From 1982 to 1986 wheat yields from the organic farm were about the average for two neighboring conventional farms. Net output from the organic farm was less than that of the conventional farm only because the organic farmer left his field fallow every third year to grow a crop of green manure (usually alfalfa). Lower expenses for fertilizer and pesticides compensated him for the lower net yield. More important, the productivity of his farm did not decline over time.

Reganold's team found that the topsoil on the organic farm was about six inches thicker than the soil of the conventional farm. The organic farm's soil had greater moisture-holding capacity and more biologically available nitrogen and potassium. Soil on the organic farm also contained many more microbes than the conventional farm's soil. Topsoil on the organic farm had more than half again as much organic matter as topsoil on the conventional farm.

The organic fields not only eroded slower than the soil replacement rates estimated by the Soil Conservation Service, the organic farm was building soil. In contrast, the conventionally farmed field shed more than six inches of topsoil between 1948 and 1985. Direct measurements of sediment yield confirmed a fourfold difference in soil loss between the two farms.

The bottom line was simple. The organic farm retained its fertility despite intensive agriculture. Soil on the conventional farm—and by implication most neighboring farms—gradually lost productive capacity as the soil thinned. With fifty more years of conventional farming, the region's topsoil will be gone. Harvests from the region are projected to drop by half once topsoil erosion leaves conventional farmers plowing the clayey subsoil. To sustain crop production, technologically driven increases in crop yields will have to double just to stay even.

European researchers also report that organic farms are more efficient and less detrimental to soil fertility. A twenty-one-year comparison of crop yields and soil fertility showed that organic plots yielded about 20 percent less than plots cultivated using pesticide-and-fertilizer-intensive methods. However, the organic plots used a third to half the input of fertilizer and energy and virtually no pesticides. In addition, the organic plots harbored far more pest-eating organisms and supported greater overall biological activity. In the organic plots, the biomass of earthworms was up to three times higher and the total length of plant roots colonized by beneficial soil mycorrhizae was 40 percent greater. Organic farming methods not only increased soil fertility, profits from organic farms were comparable to those of conventional farms. Commercially viable, organic farming need not remain an alternative philosophy.

Other recent studies support this view. A comparison of neighboring farms using organic and conventional methods on identical soils in New Zealand found the organic farms had better soil quality, higher soil organic matter, and more earthworms—and were as financially viable per hectare. A comparison of apple orchards in Washington State found similar crop yields between conventional and organic farming systems. The five-year study found that organic methods not only used less energy, maintained higher soil quality, and produced sweeter apples, they proved more profitable than conventional methods. An orchard grown under conventional methods that became profitable in about fifteen years would show profits within a decade under organic methods.

While the organic sector is the fastest-growing segment of the U.S. food market, many currently profitable conventional farming methods would become uneconomical if their true costs were incorporated into market pricing. Direct financial subsidies, and failure to include costs of depleting soil fertility and exporting pollutants, continue to encourage practices that degrade the land. In particular, the economics and practicalities of large-scale farming often foster topsoil loss and compensate with fertilizers and soil amendments. Organic farming uses fewer chemicals and—for that very reason—receives fewer research dollars per acre of production. At this point, individuals seeking healthier food contribute more to agricultural reform than do governments responsible for maintaining long-term agricultural capacity.

Over the past decade American farm subsidies averaged more than $10 billion annually. Although subsidy programs were originally intended to support struggling family farms and ensure a stable food supply, by the 1960s farm subsidies actively encouraged larger farms and more intensive methods of crop production focused on growing single crops. U.S. commodity programs that favor wheat, corn, and cotton create incentives for farmers to buy up more land and plant only those crops. In the 1970s and 1980s, subsidies represented almost a third of U.S. farm income. A tenth of the agricultural producers (coincidentally, the largest farms) now receive two-thirds of the subsidies. Critics of the subsidy program, including Nebraska Republican senator Chuck Hagel, maintain that it favors large corporate farms and does little for family farms. Good public policy would use public funds to encourage soil stewardship—and family farms, they argue—instead of encouraging large-scale monoculture.

Organic agriculture is starting to lose its status as a fringe movement as farmers relearn that maintaining soil health is essential for sustaining high crop yields. A growing shift away from agrochemical methods coincides with the renewed popularity of methods to improve the soil. Today, a middle ground is evolving in which nitrogen-fixing crops grow between row crops and as cover in the off-season, and nitrate fertilizer and pesticide are used at far lower levels than on conventional farms.

The challenge facing modern agriculture is how to merge traditional agricultural knowledge with modern understanding of soil ecology to promote and sustain the intensive agriculture needed to feed the world—how to maintain an industrial society without industrial agriculture. While the use of synthetic fertilizers is not likely to be abandoned any time soon, maintaining the increased crop yields achieved over the past half century will require widespread adoption of agricultural practices that do not further diminish soil organic matter and biological activity, as well as the soil itself.

Soil conservation methods can help prevent land degradation and improve crop yields. Simple steps to retain soil productivity include straw mulching, which can triple the mass of soil biota, and application of manure, which can increase the abundance of earthworms and soil microorganisms fivefold. Depending on the particular crop and circumstances, a dollar invested in soil conservation can produce as much as three dollars' worth of increased crop yields. In addition, every dollar invested in soil and water conservation can save five to ten times that amount in costs associated with dredging rivers, building levees, and flood control in downstream areas. Although it is hard to rally and sustain political support for treating dirt like gold, American farmers are rapidly becoming world leaders in soil conservation. Because it is prohibitively expensive to put soil back on the fields once it leaves, the best, and most cost-effective strategy lies in keeping soil on the fields in the first place.

For centuries the plow defined the universal symbol of agriculture. But farmers are increasingly abandoning the plow in favor of long-shunned no-till methods and less aggressive conservation tillage—a catchall term for practices that leave at least 30 percent of the soil surface covered with crop residue. Changes in farming practices over the past several decades are revolutionizing modern agriculture, much as mechanization did a century ago—only this time, the new way of doing things conserves soil.

The idea of no-till farming is to capture the benefits of plowing without leaving soils bare and vulnerable to erosion. Instead of using a plow to turn the soil and open the ground, today's no-till farmers use disks to mix organic debris into the soil surface and chisel plows to push seeds into the ground through the organic matter leftover from prior crops, minimizing direct disturbance of the soil. Crop residue left at the ground surface acts as mulch, helping to retain moisture and retard erosion, mimicking the natural conditions under which productive soils formed in the first place.

In the 1960s almost all U.S. cropland was plowed, but over the past thirty years adoption of no-till methods has grown rapidly among North American farmers. Conservation tillage and no-till techniques were used on 33 percent of Canadian farms in 1991, and on 60 percent of Canadian croplands by 2001. Over the same period conservation tillage grew from about 25 percent to more than 33 percent of U.S. cropland, with 18 percent managed with no-till methods. By 2004, conservation tillage was practiced on about 41 percent of U.S. farmland, and no-till methods were used on 23 percent. If this rate continues, no-till methods would be adopted on the majority of American farms in little more than a decade. Still, only about 5 percent of the world's farmland is worked with no-till methods. What happens on the rest may well shape the course of civilization.

No-till farming is very effective at reducing soil erosion; leaving the ground covered with organic debris can bring soil erosion rates down close to soil production rates—with little to no loss in crop yields. In the late 1970s, one of the first tests of the effect of no-till methods in Indiana reported a more than 75 percent reduction in soil erosion from cornfields. More recently, researchers at the University of Tennessee found that no-till farming reduced soil erosion by more than 90 percent over conventional tobacco cultivation. Comparison of soil loss from cotton fields in northern Alabama found that no-till plots averaged two to nine times less soil loss than conventional-till plots. One study in Kentucky reported that no-till methods decreased soil erosion by an astounding 98 percent. While the effect on erosion rates depends on a number of local factors, such as the type of soil and the crop, in general a 10 percent increase in ground surface cover reduces erosion by 20 percent, such that leaving 30 percent of the ground covered reduces erosion by more than 50 percent.

Lower erosion rates alone do not explain the rapid rise in no-till agri-culture's popularity. No-till methods have been adopted primarily because of economic benefits to farmers. The Food Security Acts of 1985 and 1990 required farmers to adopt soil conservation plans based on conservation tillage for highly erodible land as a condition for participating in popular USDA programs (like farm subsidies). But conservation tillage has proven to be so cost-effective that it also is being widely adopted on less erodible fields. Not plowing can cut fuel use by half, enough to more than offset income lost to reduced crop yields, translating into higher profits. It also increases soil quality, organic matter, and biota; even earthworm popula-tions are higher under no-till methods. Although adopting no-till practices can initially result in increased herbicide and pesticide use, the need declines as soil biota rebound. Growing experience in combining no-till methods with the use of cover crops, green manures, and biological pest management suggests that these so-called alternative methods offer practi-cal complements to no-till methods. Farmers are adopting no-till methods because they can both save money and invest in their future, as increasing soil organic matter means more fertile fields—and eventually lower outlays for fertilizer. The lower cost of low-till methods is fueling growing interest even among large farming operations.

No-till agriculture has another advantage; it could provide one of the few relatively rapid responses to help hold off global warming. When soil is plowed and exposed to the air, oxidation of organic matter releases car-bon dioxide gas. No-till agriculture has the potential to increase the organic matter content of the top few inches of soil by about 1 percent a decade. This may sound like a small number, but over twenty to thirty years that can add up to 10 tons of carbon per acre. As agriculture mechanized over the past century and a half, U.S. soils are estimated to have lost about 4 billion metric tons of carbon into the atmosphere. Worldwide, about 78 billion metric tons of carbon once held as soil organic matter have been lost to the atmosphere. A third of the total carbon dioxide buildup in the atmosphere since the industrial revolution has come not from fossil fuels but from degradation of soil organic matter.

Improvement of agricultural soils presents an opportunity to sequester large amounts of carbon dioxide to slow global warming—and help feed a growing population. If every farmer in the United States were to adopt no-till practices and plant cover crops, American agriculture could squirrel away as much as 300 million tons of carbon in the soil each year, turning farms into net carbon sinks, rather than sources of greenhouse gases. While this would not solve the problem of global warming—the soil can hold only so much carbon—increasing soil carbon would help buy time to deal with the root of the problem. Adoption of no-till practices on the world's 1.5 billion hectares of cultivated land has been estimated to be capable of absorbing more than 90 percent of global carbon emissions for the several decades it would take to rebuild soil organic matter. A more realistic scenario estimates the total carbon sequestration potential for the world's cropland as roughly 25 percent of current carbon emissions. Moreover, more carbon in the soil would help reduce demand for fertilizers and would lead to less erosion, and therefore further slow carbon emissions, all while increasing soil fertility.

The attraction of no-till methods is immense, but obstacles remain to their universal adoption. And they don't work well everywhere. No-till methods work best in well-drained sandy and silty soils; they do not work well in poorly drained heavy clay soils that can become compacted unless tilled. Slow-to-change attitudes and perceptions among farmers are primary factors limiting their wider adoption in the United States; the lag in adoption of no-till methods in Africa and Asia additionally reflects lack of financial resources and governmental support. In particular, small-scale farmers often lack access to specialized seed drills to plant through crop residues. Many subsistence farmers use the residue from the previous year's crops as fuel or animal fodder. These challenges are substantial, but they are well worth tackling. Reinvesting in nature's capital by rebuilding organic-rich soils may well hold the key to humanity's future.

It is no secret that if agriculture doesn't become sustainable nothing else will; even so, some still treat our soil like dirt—and sometimes worse. The eastern Washington town of Quincy is an unlikely place to uncover one of our nation's dirtiest secrets. But in the early 1990s the town's mayor clued *Seattle Times* reporter Duff Wilson in to how toxic waste was being recycled into fertilizer and sprayed on croplands. Patty Martin was an unusual candidate for a whistle blower, a conservative housewife and former pro basketball player who won a virtually uncontested race for mayor of her small farming community. When Martin's constituents began complaining about mysteriously withered crops and crop dusters spraying fertilizer out on the open prairie for no apparent reason, she learned that Cenex, a fertilizer-specialty division of the Land O'Lakes Company (yes, the butter people), was shipping toxic waste to her town, mixing it with other chemicals in a big concrete pond near the train station, and then selling the concoction as cheap, low-grade fertilizer.

It was a great scheme. Industrial polluters needing to dispose of toxic waste avoided the high cost of legitimate dumps. (Anyone who puts something into a registered toxic waste dump owns it forever.) But mixing the same stuff into cheap fertilizer and spreading it on vacant land—or selling it to farmers—makes the problem, and the liability, disappear. So trains pulled in and out of Quincy in the middle of the night and the pond went up and down with no records of what went in or out of it. Sometimes Cenex sold the new-fangled fertilizer to unsuspecting farmers. Sometimes the company paid farmers to use it just to get rid of the stuff.

Martin discovered that state officials allowed recycling waste rich in heavy metals into fertilizer without telling farmers about all those extra "nutrients." Whether or not something was considered hazardous depended not so much on what the stuff was as on what one intended to do with it. Approached about the practice of selling toxic waste as fertilizer, staff at the state department of agriculture admitted they thought it was a good idea, kind of like recycling.

Curiously enough, the toxic fertilizer began killing crops. Unless they are eroded away, heavy metals stick around in the soil for thousands of years. And if they build up enough in the soil, they are taken up by plants—like crops.

Why would a company like Cenex be mixing up a toxic brew and selling it as low-grade fertilizer? Try the oldest reason around—money. Company memos reveal that they saved $170,000 a year by calling their rinse pond waste a product and spraying it on farmer's fields. The legal case ended in 1995 when the company agreed to plead guilty to using pesticide for an unapproved purpose and pay a $10,000 fine. Now I don't particularly like to gamble, but even I'd take my savings to Vegas anytime with a guaranteed 17 to 1 payout.

After the Cenex case, other farmers in the area began to wonder whether bad fertilizer had been the cause of their failing crops. One told Martin's friend Dennis DeYoung about a fertilizer tank that Cenex delivered to his farm and forgot about years earlier. Dennis scooped out some of the dried residue from the abandoned tank and sent it off to a soil-testing lab in Idaho. The lab found lots of arsenic, lead, titanium, and chromium—not exactly premium plant food. The lab also reported high lead and arsenic concentrations in peas, beans, and potatoes DeYoung sent in from crops fertilized by Cenex products. Samples of potatoes another friend of DeYoung's sent in were found to have ten times the allowable concentration of lead.

Washington wasn't the only place where toxic waste was being reclassified as fertilizer. Between 1984 and 1992, an Oregon subsidiary of ALCOA (the Aluminum Company of America) recycled more than two hundred thousand tons of smelter waste into fertilizer. ALCOA saved two million dollars a year turning waste into a product marketed as a road de-icer during the winter and plant food in the summer. Companies all across America were saving millions of dollars a year selling industrial waste instead of paying to send it to toxic waste dumps. By the late 1990s, eight major U.S. companies converted 120 million pounds of hazardous waste into fertilizer each year.

Strangely, nobody involved seemed too anxious to talk about the toxic waste-into-fertilizer industry. They didn't have to worry. No rules prevented mixing hazardous waste into fertilizer, and then into the soil. No one appeared too concerned about such blatant disregard of the importance of healthy soil. Never mind that, as seems obvious, farms are about the last place we should use as a dumping ground for heavy metals.

The way we treat our agricultural soils, whether as locally adapted ecosystems, chemical warehouses, or toxic dumps, will shape humanity's options in the next century. Europe broke free from the ancient struggle to provide enough food to keep up with a growing population by coming to control a disproportionate share of the world's resources. The United States escaped the same cycle by expanding westward. Now with a shrinking base of arable land, and facing the end of cheap oil, the world needs new models for how to feed everybody. Island societies provide one place to look;

some consumed their future and descended into brutal competition for arable land, others managed to sustain peaceful communities. The key difference appears to be how social systems adapted to the reality of sustaining agricultural productivity without access to fresh land—in other words, how people treated their soil.

NINE

Islands in Time

When our soils are gone, we too, must go unless
we find some way to feed on raw rock.

THOMAS C. CHAMBERLAIN

ON HIS WAY TO INDONESIA AND THE SPICE ISLANDS, a Dutch admiral discovered a small volcanic island in a remote part of the Pacific Ocean on Easter Sunday 1722. Shocked by apparent cannibalism among the natives, Jakob Roggeveen and his crew barely paused and sailed on across the Pacific. Never attractive for colonization or trade because of its meager resource base, Easter Island was left alone until the Spanish annexed it half a century later. The most interesting thing about the place was a curious collection of hundreds of colossal stone heads littered across the island.

Easter Island presented a world-class puzzle to Europeans who wondered how a few stranded cannibals could have erected all those massive heads. The question mystified visitors until archaeologists pieced together the environmental history of the island to learn how a sophisticated society descended into barbarism. Today Easter Island's story provides a striking historical parable of how environmental degradation can destroy a society.

The tale is not one of catastrophic collapse but of decay occurring over generations as people destroyed their resource base. Easter Island's native civilization did not disappear overnight. It eroded away as environmental degradation reduced the number of people the island could support to fewer than those living there already. Hardly cataclysmic, the outcome was devastating nonetheless.

Pollen preserved in lake sediments records an extensive forest cover when a few dozen people colonized Easter Island. The conventional story is that Polynesians arrived in the fifth century and over the next thousand years cleared the forest for agriculture, fuel, and canoes as the population grew to almost ten thousand in the fifteenth century. Then, within a century of the peak in population, a timber shortage began forcing people to live in caves. Although recent reanalysis of radiocarbon dating suggests colonization may have occurred centuries later, pollen and charcoal from sediment cores indicate the island retained some forest cover through the seventeenth century. The island was virtually treeless by the time the first Europeans arrived. By then the last trees lay out of reach, sheltered at the bottom of the island's deepest extinct volcano.

Soil erosion accelerated once forest clearing laid the land bare. Crop yields began to fall. Fishing became more difficult after the loss of the native palms whose fibers had been used to make nets. As access to food decreased, the islanders built defensive stone enclosures for their chickens—the last food source on the island not directly affected by loss of trees and topsoil. Without the ability to make canoes, they were trapped, reduced to perpetual warfare over a diminishing resource base that ultimately came to include themselves as their society unraveled.

Rapa Nui (the native inhabitants' name for Easter Island) is located at the same latitude as central Florida, but in the Southern Hemisphere. Continually swept by warm Pacific winds, the island consists of three ancient volcanoes occupying less than fifty square miles—a tropical paradise more than a thousand miles from the nearest inhabitable land. Such isolation meant that the island supported few native plants and animals when wayward Polynesians landed after paddling across the Pacific Ocean. The native flora and fauna offered so little to eat that the new arrivals' diet was based on chickens and sweet potatoes they brought with them. Sweet potato cultivation took little effort in the island's hot, humid environment, leaving the islanders with enough free time to develop a complex society centered on carving and erecting gigantic stone heads.

The monstrous statues were carved at a quarry, transported across the island, and then capped by a massive topknot of red stone from a different quarry. The purpose of the statues remains a mystery; how the islanders did it was as much of a mystery for many years. That they transported their immense statues without mechanical devices and using only human power perplexed Europeans viewing the treeless landscape.

When asked how the great stone statues had been transported, the few remaining islanders did not know how their ancestors had done it. They simply replied that the statues walked across the island. For centuries the bare landscape fueled the mystery of the heads. No one, including the sculptors' descendants, imagined that the great stone statues were rolled on logs—it seemed just as likely that they had walked across the island on their own.

Many of the statues were left either unfinished or abandoned near their quarry, implying that their sculptors ignored the impending timber shortage until the very end. As timber became scarce, competition for status and prestige continued to motivate the drive to erect statues. Even though the Easter Islanders knew they were isolated on a world they could walk around in a day or two, cultural imperatives apparently overcame any concern about running out of trees.

European contact finished off what was left of the native culture. In the 1850s most of the island's remaining able-bodied men, including the king and his son, were enslaved and shipped off to Peruvian guano mines. Years later, the fifteen surviving abductees repatriated to their island introduced smallpox to a population with no immunity. Soon thereafter the island's population dropped to just one hundred and eleven, unraveling any remaining cultural continuity.

The story of how Easter Islanders committed ecological suicide is preserved in the island's soil. Derived from weathered volcanic bedrock, thin poorly developed soil, in places only a few inches thick, blankets most of the island. Just as in other subtropical regions, the thin topsoil held most of the available nutrients. Soil fertility declined rapidly once vegetation clearing allowed runoff to carry away the topsoil. After that only a small part of the island remained cultivatable.

Distinctively abbreviated subsoil exposed at the ground surface testifies to erosion of the island's most productive soil. Exposures at the foot of hillslopes reveal that a layer of material brought down from higher on the slopes covers the eroded remnants of the older original soil. These truncated soil profiles are studded with telltale casts of the roots of the now extinct Easter Island palm.

The relationship of soil horizons to archaeological sites reveals that most of the soil erosion occurred after construction of stone dwellings (ahus) associated with the rise of agriculture on the island. These dwellings were built directly on top of the native soil, and younger deposits of material washed off the slopes now bury the *ahus'* foundations. So the erosion that stripped topsoil from the slopes happened after the *ahus* were built.

Radiocarbon dating of the slope-wash deposits and soil profiles exposed by erosion, in road cuts, or in hand-dug soil pits record that the top of the island's original soil eroded off between about AD 1200 and 1650. Apparently, vegetation clearing for agriculture triggered widespread erosion of the A horizon upon which soil fertility depended. Easter Island's society faded soon after its topsoil disappeared, less than a century before Admiral Roggeveen's unplanned visit.

A detailed study of the soils on the Poike Peninsula revealed a direct link between changing agricultural practices and soil erosion on Easter Island. Remnants of the original soil still standing on a few tiny hills, flat-topped scraps of the original ground surface, attest to widespread erosion of the native topsoil. Downhill from these remnant soil pedestals, hundreds of thin layers of dirt, each less than half an inch thick, were deposited on top of a cultivated soil studded with the roots of the endemic palm tree. A half-inch thick layer of charcoal immediately above the buried soil attests to extensive forest clearing after a long history of cultivating plots interspersed among the palm trees.

Initial agricultural plots in planting pits dug between the trees protected the ground from strong winds and heavy rainfall, and shielded crops from the tropical sun. Radiocarbon dating of the charcoal layer and material obtained from the overlying layers of sediment indicate that the soil eroded off the upper slopes, and buried the lower slopes, between AD 1280 and 1400. The numerous individual layers of sediment deposited on the lower slopes show that the soil was stripped off storm-by-storm a fraction of an inch at a time. These observations tell the story of how after centuries of little erosion from fields tucked beneath a forest canopy, the forest of the Poike Peninsula was burned and cleared for more intensive agriculture that exposed the soil to accelerated erosion. Agriculture ceased before AD 1500, after just a century or two in which the soil slowly disappeared as runoff from each storm removed just a little more dirt.

The island's birds disappeared too. More than twenty species of seabirds inhabited Easter Island when Polynesians arrived. Just two species survived until historic times. Nesting in the island's closed canopy native forest, these birds fertilized the soil with their guano, bringing marine nutrients ashore to enrich naturally poor volcanic soils. Wiping out the island's native birds eliminated a key source of soil fertility, contributing to the decline of the soil and perhaps even the failure of the forest to regenerate.

I doubt the Easter Islanders had any idea that eating all the birds could undermine their ability to grow sweet potatoes.

The story of Easter Island is by no means unique. Catastrophic erosion followed forest clearing by Polynesian farmers on many other—but by no means all—Pacific Islands. Among the last places colonized on earth, South Pacific islands provide relatively simple settings to study the evolution of human societies because they had no land vertebrates before people imported their own fauna of chickens, pigs, dogs, and rats.

The islands of Mangaia and Tikopia provide stark contrasts in human adaptation to the realities of a finite resource base. Sharing many common traits and similar environmental histories until well after people arrived, these societies addressed declining resource abundance in very different ways. As worked out by UC Berkeley anthropologist Patrick Kirch, their stories show how transgenerational trends shaped the fate of entire societies.

Mangaia occupies just twenty square miles—a small dot of land in the South Pacific twenty-one and a half degrees south of the equator. Visited by Captain James Cook in 1777, Mangaia looks like a medieval walled fortress rising from the sea. The deeply weathered basaltic hills of the island's interior climb more than five hundred feet above sea level, surrounded by a gray coral reef lifted out of the ocean. A hundred thousand years ago, growth of the nearby volcanic island of Rarotonga warped Earth's crust enough to pop Mangaia and its fringing reef up out of the sea. Streams flowing off the island's core run into this half-mile-wide wall of razor-sharp coral that rises half the island's height. There they drop their sediment load and sink into caves running down to the island's narrow beach. Radiocarbon-dated sediment cores recovered from the base of the island's interior cliffs tell the story of Mangaia's last seven thousand years.

Covered by forest for five thousand years before Polynesians arrived about 500 BC, Mangaia eroded slowly enough to build up a thick soil in the island's volcanic core. Kirch's sediment cores record sweeping changes between 400 BC and AD 400, when a rapid increase in the abundance of microscopic charcoal particles records the expansion of slash-and-burn agriculture. Charcoal is virtually absent from sediment older than 2,400 years; dirt deposited less than 2,000 years ago contains millions of tiny carbon fragments per cubic inch of dirt. In the sediment cores, sharp increases in the abundance of iron and aluminum oxides, along with decreased phosphorus content, show that erosion of a thin, nutrient-rich layer of topsoil rapidly exposed nutrient-poor subsoil. The native forest depended on recycling nutrients that the weathered bedrock could not readily resupply. So topsoil loss retarded forest regeneration. Well adapted to grow on the nutrient-poor subsoil, ferns and scrub vegetation useless for human subsistence now cover more than a quarter of the island.

By about AD 1200 the pattern of shifting slash-and-burn agriculture had stripped so much topsoil from cultivated slopes that Mangaian agriculture shifted to reliance on labor-intensive irrigation of taro fields in the alluvial valley bottoms. Occupying just a few percent of the island's surface area, these fertile bottomlands became strategic objectives in perpetual inter-tribal warfare. Control of the last fertile soil defined political and military power on the island as population centers grew around these productive oases.

Polynesian colonization changed the ecological makeup of the island, and not only in terms of the soil. Between AD 1000 and 1650 guano-producing fruit bats vanished as the islanders killed off more than half the native bird species. Historical accounts and changes in the abundance and variety of bones in prehistoric deposits indicate that by the time of Cook's visit Mangaians had eaten all their pigs and dogs, and probably all their chickens too. The Mangaian diet began to change radically—and not for the better.

After most protein sources were gone, charred rat bones became preva-lent in deposits excavated from prehistoric rock shelters. Early nineteenth-century missionary John Williams wrote that rats were a favorite staple on Mangaia. "The natives said they were exceedingly 'sweet and good'; indeed a common expression with them, when speaking of any thing delicious, was, 'It is as sweet as a rat.'"[1] Charred, fractured, and gnawed human bones appear in excavated rock shelter deposits around AD 1500, attesting to intense competition for resources just a few hundred years before Euro-pean contact. Chronic warfare, rule by force, and a culture of terror char-acterized the end state of precontact Mangaian society.

Reconstructions of Mangaia's human population mirror those of Easter Island, albeit on a smaller scale. Starting with perhaps a few dozen colo-nizers around 500 BC, the island's population grew steadily to about five thousand people by AD 1500. The population fell dramatically over the next two centuries, hitting a low soon after European contact and then rebounding to a modern population of several thousand.

The environmental and cultural history of Tikopia, a British protec-torate in the Solomon Islands, provides a striking contrast to Mangaia despite very similar backgrounds. With a total area of less than two square miles, Tikopia is smaller than Mangaia. Even so, the two islands supported comparable populations at the time of European contact. With a population density five times greater, Tikopia sustained a relatively stable and peaceful society for well over a thousand years. This tiny island offers a model for sustainable agriculture and an encouraging example of cultural adaptation to limited resources.

Land use on Tikopia began much as that on Mangaia did. After people arrived about 900 BC, a shifting pattern of forest clearing, burning, and cultivation increased erosion rates and began to deplete the island's native fauna. After seven centuries on the island, the islanders intensified pig production, apparently to compensate for loss of birds, mollusks, and fish. Then instead of following the path taken by the Mangaians and Easter Islanders, Tikopians adopted a very different approach.

In their second millennium on their island, Tikopians began adapting their agricultural strategy. Plant remains found in the island's sediments record the introduction of tree crops. A decline in the abundance of microscopic charcoal records the end of agricultural burning. Over many generations, Tikopians turned their world into a giant garden with an overstory of coconut and breadfruit trees and an understory of yams and giant swamp taro. Around the end of the sixteenth century, the island's chiefs banished pigs from their world because they damaged the all-important gardens.

In addition to their islandwide system of multistory orchards and fields, social adaptations sustained the Tikopian economy. Most important, the islanders' religious ideology preached zero population growth. Under a council of chiefs who monitored the balance between the human population and natural resources, Tikopians practiced draconian population control based on celibacy, contraception, abortion, and infanticide, as well as forced (and almost certainly suicidal) emigration.

Arrival of Western missionaries upset the balance between Tikopia's human population and its food supply. In just two decades the island's population shot up by 40 percent after missionaries outlawed traditional population controls. When cyclones wiped out half the island's crops in two successive years, only a massive relief effort prevented famine. Afterward, the islanders restored the policy of zero population growth, this time based on the more Western practice of sending settlers off to colonize other islands.

Why did Tikopians follow such a radically different path than their counterparts on Mangaia and Easter Island? Despite similar settings and natural resources, the societies that colonized these islands met radically different fates. Tikopia developed into an idyllic island paradise, while Mangaia and Easter Island descended into perpetual warfare. Recalling that Tikopia's utopian system was maintained at the cost of lives prevented or eclipsed in the name of population control, we can justifiably ask which was the higher price. Nonetheless, Tikopian society prospered for thousands of years on a tiny isolated outpost.

An essential difference between the stories of these islands lies in their soils. Deeply weathered soils in Mangaia's sloping volcanic core are nutrient poor. The sharp coral slopes of the uplifted reef hold no soil at all. In contrast, Tikopia hosts young phosphorous-rich volcanic soils. The greater natural resilience of Tikopia's soils—because of rapid weathering of rocks with high nutrient content—enabled Tikopians to sustain key soil nutrients, using them at about the rate that they were replaced from the underlying rocks through intensive, multistory gardening that protected topsoil.

After deciphering the environmental history of both Tikopia and Mangaia, Patrick Kirch suspects that geographic scale also influenced the social choices that shaped these island societies. Tikopia was small enough that everyone knew everyone else. Kirch suggests that the fact that there were no strangers on the island encouraged collective decision making. By contrast, he suggests, Mangaia was just large enough to foster an us versus them dynamic that fueled competition and warfare between people living in neighboring valleys. Easter Island supported a larger and less cohesive society, leading to even more disastrous results. If Kirch is right and larger social systems encourage violent competition over collective compromise, we need to take a sober look at our global prospects for managing our island in space.

The story of dramatic soil loss following human colonization of islands is not restricted to the South Pacific. Viking colonization of Iceland in AD 874 catalyzed an episode of catastrophic soil erosion that continues to consume the country. At first the new colony prospered raising cattle and growing wheat. The population rose to almost eighty thousand people by AD 1100. Yet by the late eighteenth century the island's population had dwindled to half the medieval population. Cooling during the Little Ice Age from about AD 1500 to 1900 certainly influenced the fortunes of the Iceland colony. So did soil erosion.

Iceland had an extensive forest cover when first colonized. In compiling the *Íslendingabók* in the late twelfth century, Ári the Wise described the island as "forested from mountain to sea shore."[2] Since human settlement, more than half of Iceland's vegetation cover has been removed. The native birch forest that covered thousands of square miles now occupies less than 3 percent of its original area.

Over time, herds of sheep increasingly disturbed the landscape. By the start of the eighteenth century more than a quarter of a million sheep roamed the Icelandic countryside. Their numbers more than doubled by the nineteenth century. Visitors began describing Iceland as a bare land devoid of trees. The combination of a deteriorating climate and extensive overgrazing led to severe erosion and abandoned farms. Today, three-quarters of Iceland's forty thousand square miles of land are adversely affected by soil erosion; seven thousand square miles are so severely eroded as to be useless.

Once Iceland's slopes were deforested, strong winds blowing off its central ice caps helped strip the soil from roughly half the once forested area of the island. Large herds of sheep broke up the soil, allowing wind and rain to dig their way down to bedrock last exposed by melting glaciers. Soils built up over thousands of years disappeared within centuries. The central part of the island where the soil has been completely removed is now a barren desert where nothing grows and no one lives.

Some areas eroded soon after the Vikings arrived. During the relatively warm period in the eleventh and twelfth centuries, before the Little Ice Age, severe soil erosion caused the abandonment of mostly inland and some coastal farm sites. Later erosion in the lowlands primarily involved farms in marginal locations.

Many theories have been advanced to explain Iceland's abandoned farms. Inland areas have been vacated for centuries, some valleys literally deserted. Until recently, the abandonment was primarily attributed to climate deterioration and associated epidemics. But recent studies have documented the role of severe soil erosion in converting farms and grazing land into barren zones. The history of Icelandic soils can be read through the layers of volcanic ash. Frequent volcanic eruptions imprinted Iceland's dirt with a geologic bar code. Each ash buried the soils onto which they fell. The layers gradually became incorporated into the soil as wind deposited more dirt on top.

In 1638 Bishop Gisli Oddson described layers of volcanic ash in Icelandic soils. The observant bishop noticed that thick layers of ash separated buried soils, some of which contained the rooted stumps of ancient trees. Since Oddson's day, it has been recognized that hundreds of volcanic eruptions after the last glaciation produced fine-grained soils readily eroded if exposed to high winds sweeping across the island. Windblown material accumulates where vegetation stabilizes the ground surface, combining with layers of volcanic ash to build Icelandic soils. Based on ages of the different layers of ash in soil profiles, Icelandic soils accumulated at about half a foot every thousand years, roughly half an inch per century. The loss of vegetation not only accelerates erosion, but keeps soil from accumulating once there is nothing on the surface to trap volcanic ash and windblown silt.

In prehistoric times, relatively loose soil held together by thick native vegetation slowly built up on top of more cohesive lava and glacial till (an unstratified mix of clay, sand, and boulders deposited by glaciers). In areas where the soil sits directly on top of the till, soil accumulated continuously over ten thousand years. In some areas, exposed layers of soil and ash preserve evidence for erosion before the Vikings arrived, during periods when climatic deterioration stressed Iceland's native vegetation. The combination of overgrazing and climatic deterioration during the Little Ice Age triggered the most extensive episode of soil erosion in Iceland's postglacial history.

During the light-filled Icelandic summer, sheep graze twenty-four hours a day, roaming over both heath and wetlands. Trampling generates bare spots up to several feet in diameter. Shorn of a dense root mat, Iceland's volcanic soils offer little resistance to wind, rain, or snowmelt. Patches of bare earth erode rapidly down to hard rock or glacial till, carving little cliffs ranging in height from one to almost ten feet, depending on the local depth of the soil. Once started, these miniature escarpments sweep across the landscape eating away at the remaining pillars of soil and transforming rich grazing lands into windswept plains of volcanic tephra and rock fragments. Soil erosion since Norse settlement removed the original soil from about half the island. Although many factors contribute, overgrazing by sheep is generally acknowledged as the primary cause. Worms may have shaped Darwin's England (once glaciers got through with it), but sheep shaped Iceland.

Rofabards—the Icelandic name for soil escarpments—erode back half an inch to a foot and a half per year. On average, rofabard advance amounts to an annual loss of 0.2 to 0.5 percent of the soil cover from areas across which rofabards presently occur. At this rate it would take just a few hundred more years to finish stripping the soil from the whole island. Since Viking settlement, rofabard erosion has removed the soil from about five square miles per year. Icelandic scientists fear that many areas of the country have already passed a threshold that makes further erosion inevitable. They also know that once stripped of soil the land is pretty much useless.

Figure 25. Professor Ulf Helldén standing on top of a *rofabard*, the last remnant of soil that formerly covered the surrounding plain, Iceland (courtesy of Professor Helldén, Lund University).

Even though Iceland has lost 60 percent of its vegetative cover and 96 percent of its tree cover, after 1,100 years of inhabitation most Icelanders find it difficult to conceive of their modern desert as having once been forested. Most don't comprehend how severely their landscape has been degraded. Just as at Easter Island, people's conception of what is normal evolves along with the land—if the changes occur slowly enough.

The Caribbean islands of Haiti and Cuba provide another dramatic contrast in how island nations treat their soil. Haiti, which means "green island" in the native language, Arawak, is a modern example of how land degradation can bring a country to its knees. Cuba provides an example of a nation that, out of necessity, transformed a conventional agricultural system into a model for feeding a post-petroleum world.

The history of Haiti, the western third of the island of Hispaniola, shows that small hillside farms can lead to devastating soil loss even without a disastrous hurricane. Within twenty-five years after Columbus discovered Hispaniola in 1492, Spanish settlers had annihilated the island's native

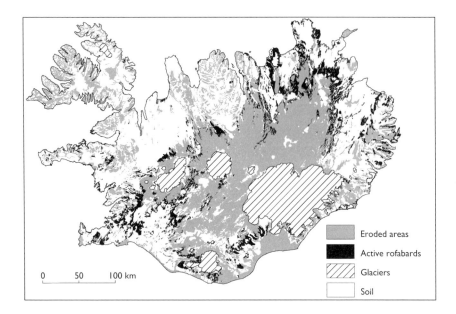

	Eroded areas
	Active rofabards
	Glaciers
	Soil

Figure 26. Map of Iceland showing the extent of areas considerably to severely eroded, glacial ice, and uneroded soils (created from data provided courtesy of Einar Grétarsson).

inhabitants. Two centuries later, in 1697, the Spanish ceded the western third of the island to the French, who imported African slaves to work timber and sugar plantations serving European markets. The colony's half million slaves revolted in the late eighteenth century, and in 1804 Haiti became the world's first republic of freed citizens to declare independence—from France, Europe's first republic.

Subsequent cultivation on steep slopes converted about a third of the country to bare rocky slopes incapable of supporting agriculture. In colonial times, there were reports of extensive erosion on upland coffee and indigo plantations and plantation owners could count on only three years of productive crops from upland fields. Widespread cultivation of steep slopes began again in the mid-twentieth century when subsistence farmers spread back into the uplands. By 1990, 98 percent of Haiti's native tropical forest was gone. Common erosion control measures such as piling up soil into mounds, or piling up soil against stakes placed along contour to create small terraces, were not very effective in controlling erosion on steep slopes.

Soil loss from the uplands in the rainy season is so severe that bulldozers function as tropical snowplows to clear the streets of the capital, Port-au-Prince. The United Nations estimates that topsoil loss over at least half the country is severe enough to preclude farming. The U.S. Agency for International Development reported in 1986 that about a third of Haiti was extremely eroded and practically sterile from soil loss. Farmers worked an area six times larger than the area well suited for cultivation. The UN Food and Agriculture Organization estimated that soil erosion destroyed 6,000 hectares of arable land a year in the 1980s. For the past few decades estimates of the remaining area of "good" farmland showed a long-term decline of several percent a year. With little more than 50 percent of the island's potential farmland still arable, the island's growing population no longer can feed itself.

Prosperity disappeared along with Haiti's topsoil. As subsistence farms literally disappeared many rural families resorted to felling the last remaining trees to sell as charcoal to buy food. Desperate peasants flocking to cities created huge slums that fostered the insurgency that toppled the government in 2004.

Haiti's crippling soil loss is not simply a colonial legacy. Land distribution in Haiti is far more egalitarian than elsewhere in Latin America. After independence the Haitian government confiscated colonial estates and freed slaves began farming unclaimed lands. Early in the nineteenth century, Haiti's president distributed a little more than 15 hectares of land to each of some ten thousand beneficiaries. Since then, land holdings generally were divided upon inheritance and several centuries of population growth gradually reduced the size of the average peasant farm to the point where by 1971, the average farm size was less than 1.5 hectares. With an average of between 5 and 6 people per household, this comes to about 0.25 to 0.3 hectares per person. More than three-fourths of rural households fall below the poverty line and two-thirds of Haitian households fall below the UN Food and Agriculture Organization's minimum nutritional standard. This is Ireland all over again, this time without the landlords.

As the population grew, the land inherited by each successive generation was subdivided into smaller plots that eventually became too small to allow fallow periods. Declining farm income reduced the ability to invest in soil conservation measures. Unable to support themselves, the poorest farmers move on to clear steeper hillsides—the only remaining land not already cultivated—and start the cycle all over on land that can last only a few years. Eventually the shortage of arable land and rising rural poverty pushes peasants from hillside subsistence farms to search for work in Port-au-Prince, where the concentration of desperate people in slums contributes to the country's tragic history of civil strife.

In Haiti, the majority of peasants own their own small farms. So small farms per se are not the answer to stopping erosion. When farms become so small that it is hard to make a living from them, it becomes hard to practice soil conservation. In Cuba, fifty miles from Haiti across the Windward Passage, the collapse of the Soviet Union set up a unique agricultural experiment. Before the 1959 Cuban revolution, the handful of people who controlled four-fifths of the land operated large export-oriented plantations, mostly growing sugar. Although small subsistence farms were still common on the remaining fifth of the land, Cuba produced less than half its own food.

After the revolution, in line with its vision of socialist progress, the new government continued sponsoring large-scale, industrial monoculture focused on export crops—primarily sugar, which accounted for three-quarters of Cuba's export income. Cuba's sugar plantations were the most mechanized agricultural operations in Latin America, more closely resembling those in California's Central Valley than on Haiti's hillsides. Farm equipment, the oil to run them, fertilizers, pesticides, and more than half of Cuba's food were imported from the island's socialist trading partners. The end of Soviet support and an ongoing U.S. trade embargo plunged Cuba into a food crisis. Unable to import food or fertilizer, Cuba saw the calories and protein in the average diet drop by almost a third, from 3,000 calories a day to 1,900 calories between 1989 and 1994.

The Soviet collapse resulted in an almost 90 percent drop in Cuba's external trade. Fertilizer and pesticide imports fell by 80 percent and oil imports fell by 50 percent. Parts to repair farm machinery were unobtainable. The *New York Times* editorial page predicted the imminent collapse of Castro's regime. Formerly one of the best-fed nations in Latin America, Cuba was not quite at the level of Haiti—but not much above it. Isolated and facing the loss of a meal a day for everyone on the island, Cuban agriculture needed to double food production using half the inputs required by conventional agriculture.

Faced with this dilemma, Cuba began a remarkable agricultural experiment, the first nation-scale test of alternative agriculture. In the mid-1980s, the Cuban government directed state-run research institutions to begin investigating alternative methods to reduce environmental impacts, improve soil fertility, and increase harvests. Within six months of the

Soviet collapse, Cuba began privatizing industrialized state farms; state-run farms were divided among former employees, creating a network of small farms. Government-sponsored farmers' markets brought peasant farmers higher profits by cutting out intermediaries. Major government programs encouraged organic agriculture and small-scale farming on vacant city lots. Lacking access to fertilizers and pesticides, the food grown in the new small private farms and thousands of tiny urban market gardens became organic not through choice but through necessity.

Charged with substituting knowledge-intensive agriculture for the embargoed inputs needed for conventional agriculture, the country's research infrastructure built on experiments in alternative agriculture that had languished under the Soviet system but were available for widespread, and immediate, implementation under the new reality.

Cuba adopted more labor-intensive methods to replace heavy machinery and chemical inputs, but Cuba's agricultural revolution was not simply a return to traditional farming. Organic farming is not that simple. You cannot just hand someone a hoe and order them to feed the proletariat. Cuba's agricultural transformation was based as much on science as was the Soviet era's high-input mechanized farming. The difference was that the conventional approach was based on applied chemistry, whereas the new approach was based on applied biology—on agroecology.

In a move pretty much the opposite of the green revolution that transformed global agriculture based on increased use of irrigation, oil, chemical fertilizers and pesticides, the Cuban government adapted agriculture to local conditions and developed biological methods of fertilization and pest control. It created a network of more than two hundred local agricultural extension offices around the country to advise farmers on low-input and no-till farming methods, as well as biological pest control.

Cuba stopped exporting sugar and began to grow its own food again. Within a decade, the Cuban diet rebounded to its former level without food imports or the use of agrochemicals. The Cuban experience shows that agroecology can form a viable basis for agriculture without industrial methods or biotechnology. Unintentionally, the U.S. trade embargo turned Cuba into a nation-scale experiment in alternative agriculture.

Some look to the Cuban example as a model for employing locally adapted ecological insight and knowledge instead of standardized mechanization and agrochemistry to feed the world. They see the solution not simply as producing cheap food, but keeping small farms—and therefore farmers—on the land, and even in cities. Thousands of commercial urban gardens grew up throughout the island, hundreds in Havana alone. Land slated for development was converted to acres of vegetable gardens that supplied markets where local people bought tomatoes, lettuce, potatoes and other crops. By 2004 Havana's formerly vacant lots produced nearly the city's entire vegetable supply.

Cuba's conversion from conventional agriculture to large-scale semi-organic farming demonstrates that such a transformation is possible—in a dictatorship isolated from global market forces. But the results are not entirely enviable; after almost two decades of this inadvertent experiment, meat and milk remain in short supply.

Cuba's labor-intensive agriculture may not produce basic crops as cheaply as American industrial farming, but the average Cuban diet did recover that lost third meal. Still, it is ironic that in retreating from the socialist agenda, this isolated island became the first modern society to adopt widespread organic and biologically intensive farming. Cuba's necessity-driven move toward agricultural self-sufficiency provides a preview of what may come on a larger scale once we burn through the supply of cheap oil that presently drives modern agriculture. And it is somewhat comforting to know that on at least one island the experiment has already been run without social collapse. Less comforting is the question of whether something similar could be pulled off in a society other than a one-party police state.

After Darwin's famous sojourn in the Galapagos, the isolated nature of islands strongly influenced biological theory. But it is only in the last several decades that such thinking reached the realm of anthropology. While people may someday migrate into space to colonize other planets, the vast majority of us remain trapped on our planet for the foreseeable future. Although a global rerun of Haiti, Mangaia, or Easter Island is by no means inevitable, the experiences of societies on islands around the world remind us that Earth is the ultimate island, an oasis in space rendered hospitable by a thin skin of soil that, once lost, rebuilds only over geologic time.

TEN

Life Span of Civilizations

Speak to the earth, and it shall teach thee.

JOB 12:8

AFTER TWO HUNDRED YEARS, THE CONTRASTING VISIONS of Malthusian pessimism and Godwinian optimism still frame debate over whether technological innovation will keep meeting society's growing agricultural needs. Preventing a substantial decline in food production once we exhaust fossil fuels will require radically restructuring agriculture to sustain soil fertility, or developing massive new sources of cheap energy if we continue to rely on chemical fertilizers. But the future is clear if we continue to erode the soil itself.

Estimating how many people Earth can support involves assumptions about trade-offs between population size, quality of life, and environmental qualities such as biodiversity. Most demographic estimates anticipate more than ten billion people on the planet by the end of this century. Whether we endorse the National Conference of Catholic Bishops' apparent belief that the world could comfortably support forty billion people, or Ted Turner's view that four hundred million would be plenty, feeding even the middle range of such estimates presents an impossible challenge. For even if we were we to somehow harness Earth's full photosynthetic production with the same efficiency as the 40 percent now devoted to supporting humanity, we could support fifteen billion people—and share the planet with nothing else.

Credible scientists also disagree on Earth's carrying capacity. Norman Borlaug, the Nobel Prize–winning green revolution pioneer, claims that Earth can support ten billion folks, although he acknowledges that it will require major advances in agricultural technology. This is the same guy who warned at his Nobel acceptance speech that the green revolution had bought us only a few decades to deal with overpopulation. Now, more than three decades later, he trusts scientists will pull more rabbits out of the hat. At the other end of the spectrum are Stanford University biologists Paul and Anne Ehrlich who maintain that we have already passed the carrying capacity of the planet, which they put at about three billion people. In their view, we've already ensured disaster.

Regardless of who is right, a key issue for any long-term scenario is reforming agriculture in both industrialized and developing countries. Conventional industrial farmers sacrifice soil to maximize short-term returns to pay rent, service debt for machinery, and buy pesticides and fertilizers. Peasant farmers mine the soil because they are trapped farming plots too small to feed their families. While the underlying economic and social issues are complex, sustaining agricultural productivity in both the developed and developing world depends on retaining fertile soil.

Irreplaceable over human timescales, soil is an awkward hybrid—an essential resource renewable only at a glacial pace. Like many environmental problems that become harder to address the longer they are neglected, soil erosion threatens the foundation of civilization over timescales longer than social institutions last. Yet as long as soil erosion continues to exceed soil production, it is only a matter of time before agriculture fails to support a growing population.

At its peak, the Roman Empire relied on slave labor to work the plantations that replaced the conservative husbandry of farmer-citizens in the early republic. Before the Civil War, the American South became addicted to similar methods that destroyed soil fertility. In both cases, soil-destroying practices became entrenched as lucrative cash crops seduced large landowners and landlords. Soil loss occurred too slowly to warrant societal attention.

There are plenty of reasons to argue for smaller, more efficient government; market efficiencies can be effective drivers for most social institutions. Agriculture is not one of them. Sustaining our collective well-being requires prioritizing society's long-term interest in soil stewardship; it is an issue of fundamental importance to our civilization. We simply cannot afford to view agriculture as just another business because the economic benefits of soil conservation can be harvested only after decades of stewardship, and the cost of soil abuse is borne by all.

The idea of free markets for labor, land, and capital developed alongside Malthus's controversial theory. Adam Smith, the father of modern economic theory, wrote his *Inquiry into the Nature and Causes of the Wealth of Nations* in 1776. In it he argued that competition between individuals acting in their own interest, whether as buyers or sellers, would produce the greatest societal benefit. Clearly, the past few centuries proved that self-regulating, free markets can effectively set prices and match production to demand. Yet even Smith acknowledged that governmental regulation is needed to steer markets toward desirable outcomes.

Almost unquestioningly accepted in Western societies, classical economics distilled from Smith's views, as well as variants like Keynesian economics, neglect the fundamental problem of resource depletion. They share the false assumption that the value of finite resources is equal to the cost of using them, extracting them, or replacing them with other resources. This problem is central to soil exhaustion and erosion, given the long time required to rebuild soil and the lack of any viable substitute for healthy soil.

Marxist economics shares this critical blind spot. Marx and Engels viewed the value of products as derived from the labor that went into their production. To them, the level of effort needed to find, extract, and use a resource accounted for issues deriving from resource scarcity. Focused on harnessing nature to advance the proletariat, they never put the idea that society could run out of key resources in their lexicon. Instead, Engels tersely dismissed the problem of soil degradation. "The productivity of the land can be infinitely increased by the application of capital, labour and science."[1] Contrary to his dour image, Engels was apparently an optimist.

In effect, economic theory—whether capitalist or Marxist—implicitly assumes that resources are inexhaustible or infinitely substitutable. Given either scenario, the most rational course of action for individuals pursing their own self-interest is to simply ignore the interests of posterity. Economic systems of all stripes are biased toward using up finite resources and passing the bill on to future generations.

Concern over long-term productivity of the soil is almost universal among those who have examined the issue. Predictably—and understandably—more pressing problems than saving dirt usually carry the day. Long-term issues seldom get addressed when more immediate crises compete for policymakers' attention. When there is lots of land, there is little incentive to preserve the soil. It is only when scarcity arrives that people notice the problem. Like a disease that remains undetected until its last stages, by then it has already become a crisis.

Just as lifestyle influences a person's life expectancy within the constraints of the human life span, the way societies treat their soil influences their longevity. Whether, and the degree to which, soil erosion exceeds soil production depends on technology, farming methods, climate, and population density. In the broadest sense, the life span of a civilization is limited by the time needed for agricultural production to occupy the available arable land and then erode through the topsoil. How long it takes to regenerate the soil in a particular climate and geologic setting defines the time required to reestablish an agricultural civilization—providing of course that the soil is allowed to rebuild.

This view implies that the life expectancy for a civilization depends on the ratio of the initial soil thickness to the net rate at which it loses soil. Studies that compare recent erosion rates to long-term geologic rates find increases of at least twofold and as much as a hundred times or more. Human activities have increased erosion rates severalfold even in areas with little apparent acceleration of erosion, while areas with acknowledged problems erode a hundred to even a thousand times faster than what is geologically normal. On average, people appear to have increased soil erosion at least tenfold across the planet.

Several years ago, University of Michigan geologist Bruce Wilkinson used the distribution and volume of sedimentary rocks to estimate rates of erosion over geologic time. He estimated that the average erosion rate over the last 500 million years was about an inch every 1,000 years, but that today it takes erosion less than 40 years, on average, to strip an inch of soil off agricultural fields—more than twenty times the geologic rate. Such dramatic acceleration of erosion rates makes soil erosion a global ecological crisis that, although less dramatic than an Ice Age or a comet impact, can prove equally catastrophic—in time.

With soil production rates of inches per millennia and soil erosion rates under conventional, plow-based agriculture of inches per decade to inches per century, it would take several hundred to a couple thousand years to erode through the one- to three-foot-thick soil profile typical of undisturbed areas of temperate and tropical latitudes. This simple estimate of the life span of civilizations predicts remarkably well the historical pattern for major civilizations around the world.

Except for the fertile river valleys along which agriculture began, civilizations generally lasted eight hundred to two thousand years, roughly thirty to seventy generations. Throughout history, societies grew and prospered as long as there was new land to plow or the soil remained productive. Things eventually fell apart when neither remained possible. Societies that prospered for longer either figured out how to conserve soil, or were blessed with an environment that naturally refreshed their dirt.

Even a casual reading of history shows that under the right circumstances any one, or any combination of political turmoil, climatic extremes, or resource abuse can bring down a society. Alarmingly, we face the potential convergence of all three in the upcoming century as shifting climate patterns and depleted oil supplies collide with accelerated soil erosion and loss of farmland. Should world fertilizer or food production falter, political stability could hardly endure.

The only ways around the boom-and-bust cycle that has characterized agricultural societies are to continuously reduce the amount of land needed to support a person, or limit population and structure agriculture so as to maintain a balance between soil production and erosion. This presents several near-term alternatives: we can fight over farmland as the human population keeps growing and soil fertility declines, maintain blind faith in our ability to keep increasing crop yields, or find a balance between soil production and erosion.

Whatever we do, our descendants will be compelled to adhere to something close to a balance—whether they want to or not. In so doing they will face the reality that agricultural reliance on fossil fuels and fertilizers parallels ancient practices that led to salinization in semiarid regions and soil loss with agricultural expansion from floodplains up into sloping terrain. Technology, whether in the form of new plows or genetically engineered crops, may keep the system growing for a while, but the longer this works the more difficult it becomes to sustain—especially if soil erosion continues to exceed soil production.

Part of the problem lies in the discrepancy between rates at which civilizations and individuals respond to stimuli. Actions that are optimal for farmers are not necessarily consistent with their societies' interests. Evolving gradually and almost imperceptibly to individual observers, the ecology of economies helps define the life span of civilizations. Societies that deplete natural stocks of critical renewable resources—like soil—sow the seeds of their own destruction by divorcing economics from a foundation in the supply of natural resources.

Small societies are particularly vulnerable to disruption of key lifelines, such as trading relations, or to large perturbations like wars or natural disasters. Larger societies, with more diverse and extensive resources, can rush aid to disaster victims. But the complexity that brings resilience may also impede adaptation and change, producing social inertia that maintains collectively destructive behavior. Consequently, large societies have difficulty adapting to slow change and remain vulnerable to problems that eat away their foundation, such as soil erosion. In contrast, small systems are adaptable to shifting baselines but are acutely vulnerable to large perturbations. But unlike the first farmer-hunter-gatherers who could move around when their soil was used up, a global civilization cannot.

In considering possible scenarios for our future, the first issue we need to consider is how much cultivatable land is available, and when we will run out of unused land. Globally about one and a half billion hectares are now in agricultural production. Feeding a doubled human population without further increasing crop yields would require doubling the area presently under cultivation. But we are already out of virgin land that could be brought into long-term production. Such vast tracts of land could be found only in tropical forests and subtropical grasslands—like the Amazon and the Sahel. Experience shows that farming such marginal lands will produce an initial return until the land quickly becomes degraded, and then abandoned—if the population has somewhere to go. Look out the plane window on a flight from New Orleans to Chicago, or Denver to Cincinnati. Everything you see is already in agricultural production. This huge expanse of naturally fertile ground literally feeds the world. The suburbs growing around any city show that we are losing agricultural land even as the human population continues to grow. With the land best suited for agriculture already under cultivation, agricultural expansion into marginal areas is more of a delaying tactic than a viable long-term strategy.

Second, we need to know how much soil it takes to support a person, and how far we can reduce that amount. In contrast to the amount of arable land, which has varied widely through time and across civilizations, the amount of land needed to feed a person has gradually decreased over recorded history. Hunting and gathering societies needed 20 to 100 hectares of land to support a person. The shifting pattern of cultivation that characterized slash-and-burn agriculture took 2 to 10 hectares of land to support a person. Later sedentary agricultural societies used about a tenth as much land to support a person. An estimated 0.5 to 1.5 hectares of floodplain fed a Mesopotamian.

Over time, human ingenuity increased food production on the most intensively farmed and productive land so that today, with roughly 6 billion people and 1.5 billion hectares of cultivated land, it takes about 0.25 hectares to feed each person. The world's most intensively farmed regions use about 0.2 hectares to support a person. Increasing the average global agricultural productivity to this level would support 7.5 billion people. Yet by 2050 the amount of available cropland is projected to drop to less than 0.1 hectare per person. Simply staying even in terms of food production will require major increases in per hectare crop yields—increases that simply may not be achievable despite human ingenuity.

Before 1950 most of the increase in global food production came from increased acreage under cultivation and improved husbandry. Since 1950 most of the increase has come from mechanization and intensified use of chemical fertilizers. Dramatic intensification of agricultural methods during the green revolution is credited with averting a food crisis over the past three decades. Increased harvests stemmed from development of high-yield "miracle" varieties of wheat and rice capable of producing two or three harvests a year, increased use of chemical fertilizers, and massive investments in irrigation infrastructure in developing nations. The introduction of fertilizer-responsive rice and wheat increased crop yields between the 1950s and 1970s by more than 2 percent a year.

Since then, however, growth in crop yields has slowed to a virtual standstill. The great postwar increase in crop yields appears to be over. Wheat yields in the United States and Mexico are no longer increasing. Asian rice yields are starting to fall. Crop yields appear to have reached a technological plateau. Thirty-year experiments on response to nitrogen fertilization at the International Rice Research Institute in the Philippines found that increasing nitrogen inputs were needed just to maintain crop yields. "At best, we have been able to keep rice yields from decreasing despite considerable investment in breeding efforts and agronomic research to improve crop management."[2] We're still waiting for the next innovation to crank up food production despite the reality that over the coming decades further annual increases of more than 1 percent are needed to meet projected demand for wheat, rice, and maize. Achieving and sustaining such increases through conventional means will require major breakthroughs as agricultural productivity approaches biological limits. It is getting harder just to stay even, let alone increase crop yields.

In the second half of the twentieth century, food production doubled thanks in great part to a sevenfold increase in nitrogen fertilization and a three-and-a-half-fold increase in phosphorus fertilization. Repeating this story simply is not possible because you can apply only so much fertilizer before plants have all they can possibly use. Even tripling fertilizer applications won't help much if soils are already saturated with biologically useful nitrogen and phosphorus. Since crops don't take up half the nitrogen in the fertilizers farmers apply today, it may not do much good to add more—even if we could.

Growing food hydroponically—by pumping water and nutrients through dirt in a laboratory—can produce far more per unit area than growing food in natural soil, but the process requires using large external inputs of nutrients and energy. This might work on small-scale, labor-intensive farms, but it cannot feed the world from large operations without huge continuous inputs of fossil fuels and nutrients mined from somewhere else.

Finally, in all likelihood the easiest—and greatest—increases in crop yields from plant breeding have already been achieved. Given a fixed gene pool already subjected to intensive natural selection over millions of years, further major gains in crop yields would require working around morphological and physiological constraints imposed by evolution. Growth in crop yields has already slowed while the cost of research to bring even incremental increases in crop production has skyrocketed. Perhaps genetic engineering might yet substantially increase crop yields—but at the risk of releasing supercompetitive species into agricultural and natural environments with unknowable consequences.

Meanwhile, global grain reserves—the amount of grain stored on hand at a given time—fell from a little more than a year's worth in 2000 to less than a quarter of annual consumption in 2002. Today the world is living harvest-to-harvest just like Chinese peasants in the 1920s. Now that's progress.

Clearly, more of the same won't work. Projecting past practices into the future offers a recipe for failure. We need a new agricultural model, a new farming philosophy. We need another agricultural revolution.

Agricultural philosopher Wendell Berry argues that economies can be based on either industrial or agrarian ideals, and that an agrarian society need not be a subsistence society lacking technological sophistication and material well-being. He sees industrial societies as based on the production and use of products, whether fundamental to survival (food) or manufactured along with the desire for it (pop tarts). In contrast, an agrarian economy is based on local adaptation of economic activity to the capacity of the land to sustain such activity. Not surprisingly, Berry likes to talk about the difference between good farming and the most profitable farming. Still, he points out that everybody need not be a farmer in an agrarian society, nor need industrial production be limited to the bare necessities. The distinction in Berry's view is that agriculture and manufacturing in an agrarian society would be tailored to the local landscape. While it is difficult to reconcile current trends with this vision for an agrarian economy, a reoriented capitalism is not unimaginable. After all, today's quasi-sovereign global corporations were inconceivable just a few centuries ago.

Agriculture has experienced several revolutions in historical times: the yeoman's revolution based on relearning Roman soil husbandry and the agrochemical and green revolutions based on fertilizer and agrotechnology. Today, the growing adoption of no-till and organic methods is fostering a modern agrarian revolution based on soil conservation. Whereas past agricultural revolutions focused on increasing crop yields, the ongoing one needs to sustain them to ensure the continuity of our modern global civilization.

The philosophical basis of the new agriculture lies in treating soil as a locally adapted biological system rather than a chemical system. Yet agroecology is not simply a return to old labor-intensive ways of farming. It is just as scientific as the latest genetically modified technologies—but based on biology and ecology rather than chemistry and genetics. Rooted in the complex interactions between soil, water, plants, animals, and microbes, agroecology depends more on understanding local conditions and context than on using standardized products or techniques. It requires farming guided by locally adapted knowledge—farming with brains rather than by habit or convenience.

Agroecology doesn't mean simply going organic. Even forgoing pesticides, California's newly industrialized organic factory farms are not necessarily conserving soil. When demand for organic produce began to skyrocket in the 1990s, industrial farms began planting monocultural stands of lettuce that retained the flaws of conventional agriculture—just without the pesticides.

Agroecology doesn't necessarily mean small farms instead of large farms. Haiti's tiny peasant farms destroyed soil on steep slopes just as effectively as the immense slave-worked plantations of the American South. And the problem isn't just mechanization. Roman oxen slowly stripped soil as effectively as the diesel-powered descendants of John Deere's plows. The underlying problem is confoundingly simple: agricultural methods that lose soil faster than it is replaced destroy societies. Fortunately, there are ways for very productive farms to operate without cashing in the soil. Put simply, we need to adapt what we do to where we do it.

Clues about how to do so may lie in the experiences of labor-intensive and technology-intensive agricultural societies. In labor-intensive systems people tend to adapt to the land. In technology-intensive systems people typically try to adapt the land to their techniques. Labor-intensive cultures that invested in their soil by increasing soil organic matter, terracing hillslopes, and recycling critical nutrients survived for long periods of time in lowland China, Tikopia, the Andes, and the Amazon. Technology-intensive societies that treated the soil as a consumable input developed systems where tenant farmers and absentee landlords extracted as much as they could from the soil as fast as possible by exchanging soil fertility for short-term profits.

This fundamental contrast highlights the problem that dirt is virtually worthless and yet invaluable. The cheapest input to agricultural systems, soil will always be discounted—until it is too late. Consequently, we need to consciously adapt agriculture to reality rather than vice versa. Human practices and traditions shaped to the land can be sustained; the opposite cannot.

Some changes in practices or habits simply require a different mind set, like no-till agriculture, which is effective at retarding soil loss and compatible with both conventional and organic agricultural practices. Nothing really stands in its way, and as experience grows it is being adopted by many U.S. farmers. For other alternative ideas, like organic practices and biological pest control, it is consumers rather than governments who are driving the process of change in today's global economy without a global society.

But governments still have an important role to play. In the developed world, through policies and subsidies they can reshape incentives to promote both small-scale organic farms and no-till practices on large, mechanized farms. In developing countries, they can give farmers new tools to replace their plows and promote no-till and organic methods on small labor-intensive farms. Governments can also support urban agriculture and much needed research on sustainable agriculture and new technologies, especially precision application of nitrogen and phosphorus, and on methods for retaining soil organic matter and soil fertility. What governments need not promote are genetic engineering and more intensive fertilizer- and irrigation-based farming—the very practices pushed by industry as the key to extending reliance on its products.

Emerging interest in supporting an agrarian land ethic is embodied in the slow food and eat-local movements that try to shorten the distance between crop production and consumption. Yet energy efficiency in the delivery of food to the table is not some radical new idea. Romans shipped grain around the Mediterranean because the wind provided the energy needed to transport food long distances. That's why North Africa, Egypt, and Syria fed Rome—it was too inefficient (and difficult) to drag western European produce over the mountains into central Italy.

Similarly, as oil becomes more expensive it will make less sense to ship food halfway around the world: the unglobalization of agriculture will become increasingly attractive and cost effective. The average piece of organic produce sold in American supermarkets travels some 1,500 miles between where it is grown and where it is consumed. Over the long run, when we consider the effect on the soil and on a post-oil world, markets for food may work better (although not necessarily more cheaply) if they are smaller and less integrated into a global economy, with local markets selling local food. As it becomes increasingly expensive to get food produced elsewhere to the people, it will become increasingly attractive to take food production to the people—into the cities.

Despite its seemingly contradictory name, urban agriculture is not an oxymoron. Throughout much of preindustrial history city wastes were primarily organic and were returned to urban and quasi-urban farms to enrich the soil. In the mid-nineteenth century, one sixth of Paris was used to produce more than enough salad greens, fruits, and vegetables to meet the city's demand—fertilized by the million tons of horse manure produced by the city's transportation system. More productive than modern industrial farms, the labor-intensive system became so well known that intensive compost-based horticulture is still called French gardening.

Urban farming has been growing rapidly—worldwide more than 800 million people are engaged in urban agriculture to some degree. The World Bank and the UN Food and Agriculture Organization encourage urban farming in efforts to feed the urban poor in developing countries. But urban farming is not restricted to developing countries; by the late 1990s one out of ten families in some U.S. cities were engaged in urban agriculture, as were two-thirds of Moscow's families. Urban farms not only deliver fresh produce to urban consumers the same day it is harvested, with lower transportation costs and the use of far less water and fertilizer, they can absorb a significant amount of solid and liquid waste, reducing urban waste disposal problems and costs. Eventually it may well be worth reconfiguring the downstream end of modern sewage systems to close the loop on nutrient cycling by returning the waste from livestock and people back to the soil. As archaic as it may sound, someday our collective well-being is likely to depend on it.

At the same time, we can't afford to lose any more farmland. Fifty years from now every hectare of agricultural land will be crucial. Every farm that gets paved over today means that the world will support fewer people down the road. In India, where we would expect farmland to be sacred, farmers near cities are selling off topsoil to make bricks for the booming housing market. Developing nations simply cannot afford to sell off their future this way, just as the developed world cannot pave its way to sustainability. Agricultural land should be viewed—and treated—as a trust held by farmers today for farmers tomorrow.

Still, farms should be owned by those who work them—by people who know their land and who have a stake in improving it. Tenant farming is not in society's best interest. Private ownership is essential; absentee landlords give little thought to safeguarding the future.

Viewed globally, humanity need not face a stark choice between eating and saving endangered species. Protecting biodiversity does not necessarily require sacrificing productive agricultural land because soils with high agricultural productivity tend to support low biodiversity. Conversely, areas with high biodiversity tend to be areas with low agricultural potential. In general, species-rich tropical latitudes tend to have nutrient-poor soils, and the world's most fertile soils are found in the species-poor loess belts of the temperate latitudes.

Much of the recent loss of biodiversity has been encouraged by government subsidies and tax incentives that allowed clearing and plowing of lands (like tropical rainforests) that can be profitably farmed for only a short period and are often abandoned once the subsidies lapse (or the soil erodes). Unfortunately, most developing countries are in the tropical latitudes where soils are both poor in nutrients and vulnerable to erosion. Despite this awkward geopolitical asymmetry, it is myopic to ignore the reality that development built upon mining soil guarantees future food shortages.

There are three great regions that could sustain intensive mechanized agriculture—the wide expanses of the world's loess belts in the American plains, Europe, and northern China, where thick blankets of easily farmed silt can sustain intensive farming even once the original soil disappears. In the thin soils over rock that characterize most of the rest of the planet, the bottom line is that we have to adapt to the capacity of the soil rather than vice versa. We have to work with the soil as an ecological rather than an industrial system—to view the soil not as a factory but as a living system. The future of humanity depends as much on this philosophical realignment as on technical advances in agrotechnology and genetic engineering.

Capital-intensive agricultural methods will never provide the third of humanity that lives on less than two dollars a day a way out of hunger and poverty. Labor-intensive agriculture, however, could—if those people had access to fertile land. Fortunately, such methods are also those that could help rebuild the planet's soil. We should be subsidizing small subsistence farmers in the developing world; teaching people how to use their land more productively invests in humanity's future. Too often, however, modern agricultural subsidies favor large industrial farms and reward farmers for practices that undermine humanity's long-term prospects.

The more than three hundred billion dollars in global agricultural subsidies amounts to more than six times the world's annual development assistance budget. Oddly, we are paying industrial farmers to practice unsustainable agriculture that undercuts the ability of the poor to feed themselves—the only possible solution to global hunger. Political systems perpetually focused on the crisis du jour rarely address chronic problems like soil erosion; yet, if our society is to survive for the long haul, our political institutions need to focus on land stewardship as a mainstream—and critical—issue.

Over the course of history, economics and absentee ownership have encouraged soil degradation—on ancient Rome's estates, nineteenth-century southern plantations, and twentieth-century industrialized farms. In all three cases, politics and economics shaped land-use patterns that favored mining soil fertility and the soil itself. The overexploitation of both renewable and nonrenewable resources is at once well known and almost impossible to address in a system that rewards individuals for maximizing the instantaneous rate of return, even if it depletes resources critical for the long term. The worldwide decimation of forests and fisheries provide obvious examples, but the ongoing loss of the soil that supplies more than 95 percent of our food is far more crucial. Other, nonmarket mechanisms—whether cultural, religious, or legal—must rise to the challenge of maintaining an industrial society with postindustrial agriculture. Counterintuitively, for the world beyond the loess belts this challenge requires more people on the land, practicing intensive organic agriculture on smaller farms, using technology but not high capitalization.

Meeting this challenge would also help address the problem of world hunger because if we are to feed the developing world, we must abandon the intuitive, but naive, idea that producing cheap food will eliminate hunger. We've already made food cheap and there are still plenty of hungry people on the planet. A different approach—one that might actually work—is to promote the prosperity of small farms in developing countries. We need to enable peasant farmers to feed themselves, and generate an income capable of lifting them out of poverty while making them stewards of the land through access to knowledge, the right tools, and enough land to both feed themselves and grow a marketable surplus.

As much as climate change, the demand for food will be a major driver of global environmental change throughout the coming decades. Over the past century, the effects of long-term soil erosion were masked by bringing new land under cultivation and developing fertilizers, pesticides, and crop varieties that compensate for declining soil productivity. However, the greatest benefits of such technological advances accrue in applications to deep, organic-rich topsoil. Agrotech fixes become progressively more difficult to maintain as soil thins because crop yields decline exponentially with soil loss. Coupled with the inevitable end of fossil-fuel-derived fertilizers, the ongoing loss of cropland and soil poses the problem of feeding a growing population from a shrinking land base. Whereas the effects of soil erosion can be temporarily offset with fertilizers and in some cases irrigation, the long-term productivity of the land cannot be maintained in the face of reduced soil organic matter, depleted soil biota, and thinning soil that so far have characterized industrial agriculture.

Many factors may contribute to ending a civilization, but an adequate supply of fertile soil is necessary to sustain one. Using up the soil and moving on to new land will not be a viable option for future generations. Will modern soil conservation efforts prove too little and too late, like those of ancient societies? Or will we relearn how to preserve agricultural soils as we use them even more intensively? Extending the life span of our civilization will require reshaping agriculture to respect the soil not as an input to an industrial process, but as the living foundation for material wealth. As odd as it may sound, civilization's survival depends on treating soil as an investment, as a valuable inheritance rather than a commodity—as something other than dirt.

NOTES

2. SKIN OF THE EARTH

1. Darwin 1881, 4.
2. Darwin 1881, 313.

3. RIVERS OF LIFE

1. Wallace 1883, 15.
2. Helms 1984, 133.
3. Lowdermilk 1926, 127, 129.

4. GRAVEYARD OF EMPIRES

1. Xenophon *Oeconomicus* 16.3.
2. Plato *Critias* 3.III.
3. Varro *De re rustica* 1.3.
4. Varro *De re rustica* 1.2.6.
5. Columella *De re rustica* 1.7.6.
6. Tacitus *Annals* 3.54.
7. Tertullian *De anima* 30.
8. Simkhovitch 1916, 209.
9. Tacitus *Annals* 2.59.

10. Marsh 1864, 9, 42.
11. Lowdermilk 1953, 16.
12. On fields, Exodus 23:10–11; and on compost, Isaiah 25:10.
13. Lowdermilk 1953, 38.
14. Cook 1949, 42.

5. LET THEM EAT COLONIES

1. Simkhovitch 1913, 400.
2. Markham 1631, 1, 3.
3. Evelyn 1679, 288–89, 295.
4. Evelyn 1679, 298, 315.
5. Mortimer 1708, 12.
6. Mortimer 1708, 14.
7. Mortimer 1708, 79.
8. Surell 1870, 135.
9. Surell 1870, 219.
10. Marsh 1864, 201.
11. Brown 1876, 10.
12. Melvin 1887, 472.
13. Hutton 1795, 205.
14. Playfair 1802, 99.
15. Playfair 1802, 106–7.
16. Godwin 1793, 2:861.
17. Marx 1867, 638.

6. WESTWARD HOE

1. Craven 1925, 19.
2. Beer 1908, 243.
3. Craven 1925, 41.
4. Hartwell, Blair, and Chilton 1727, 6, 7.
5. Phillips 1909, 1:282.
6. Eliot 1934, 223–24.
7. Hewatt 1779, 2:305, 306.
8. Brissot de Warville 1794, 1:378.
9. Craven 1833, 150.
10. Toulmin 1948, 71.
11. Washington 1803, 6.
12. Washington 1892, 13:328–29.
13. Jefferson 1894, 3:190.

14. Washington 1803, 103–4.
15. Jefferson 1813, 509.
16. Taylor 1814, 11, 15, 10.
17. de Beaujour 1814, 85–86.
18. Lorain 1825, 240.
19. Phillips 1909, 1:284–85.
20. Craven 1925, 81.
21. Letter from Alabama 1833.
22. Ruffin 1832, 15.
23. C. Lyell 1849, 2:24.
24. C. Lyell 1849, 2:36.
25. C. Lyell 1849, 2:72.
26. U.S. Senate 1850, 7–8.
27. U.S. Senate 1850, 9.
28. White 1910, 40, 45.
29. Glenn 1911, 11, 19.
30. Hall 1937, 1.

7. DUST BLOW

1. Johnson 1902, 638, 653.
2. Shaler 1905, 122–24, 128.
3. Shaler 1891, 330.
4. Shaler 1891, 332.
5. Davis 1914, 207, 213.
6. Davis 1914, 216–17.
7. Throckmorton and Compton 1938, 19–20.
8. Great Plains Committee 1936, 4.
9. Sampson 1981, 17.
10. Lowdermilk 1953, 26.
11. Schickele, Himmel, and Hurd 1935, 231.
12. National Research Council 1989, 9.
13. Ponting 1993, 261–62.

8. DIRTY BUSINESS

1. Liebig 1843, 63.
2. Hilgard 1860, 361.
3. Jenny 1961, 9–10.
4. Whitney 1909, 66.
5. van Hise 1916, 321–22.

6. USDA 1901, 31.
7. Whitney 1925, 12, 39.
8. Crookes 1900, 6, 7.
9. Smil 2001, 139.
10. Howard 1940, 4.
11. Howard 1940, 219–20.
12. Faulkner 1943, 5.
13. Faulkner 1943, 84, 127–28.

9. ISLANDS IN TIME

1. Williams 1837, 244–45.
2. Buckland and Dugmore 1991, 116.

10. LIFE SPAN OF CIVILIZATIONS

1. Engels 1844, 58.
2. Cassman et al. 1995, 218.

BIBLIOGRAPHY

I. GOOD OLD DIRT

Hooke, R. LeB. 1994. On the efficacy of humans as geomorphic agents. *GSA Today* 4:217, 224–25.

———. 2000. On the history of humans as geomorphic agents. *Geology* 28:843–46.

2. SKIN OF THE EARTH

Darwin, C. 1881. *The Formation of Vegetable Mould, Through the Action of Worms, With Observations on Their Habits.* London: John Murray.

Davidson, D. A. 2002. Bioturbation in old arable soils: Quantitative evidence from soil micromorphology. *Journal of Archaeological Science* 29:1247–53.

Gilbert, G. K. 1877. *Geology of the Henry Mountains.* U.S. Geographical and Geological Survey of the Rocky Mountain Region. Washington, DC: Government Printing Office.

Jenny, H. 1941. *Factors of Soil Formation: A System of Quantitative Pedology.* New York: McGraw-Hill.

Mitchell, J. K., and G. D. Bubenzer. 1980. Soil loss estimation. In *Soil Erosion,* ed. M. J. Kirkby and R. P. C. Morgan, 17–62. Chichester: John Wiley and Sons.

Retallack, G. J. 1986. The fossil record of soils. In *Paleosols: Their Recognition and Interpretation,* ed. V. P. Wright, 1–57. Oxford: Blackwell Scientific Publications.

Schwartzman, D. W., and T. Volk. 1989. Biotic enhancement of weathering and the habitability of Earth. *Nature* 340:457–60.

Torn, M. S., S. E. Trumbore, O. A. Chadwick, P. M. Viktousek, and D. M. Hendricks. 1997. Mineral control of soil organic carbon storage and turnover. *Nature* 389:170–73.

Wolfe, B. E., and J. N. Kilronomos. 2005. Breaking new ground: Soil communities and exotic plant invasion. *BioScience* 55:477–87.

3. RIVERS OF LIFE

Butzer, K. W. 1976. *Early Hydraulic Civilization in Egypt: A Study in Cultural Ecology.* Chicago: University of Chicago Press.

Haub, C. 1995. How many people have ever lived on Earth? *Population Today,* February.

Helms, D. 1984. Walter Lowdermilk's journey: Forester to land conservationist. *Environmental Review* 8:132–45.

Henry, D. O. 1989. *From Foraging to Agriculture: The Levant at the End of the Ice Age.* Philadelphia: University of Pennsylvania Press.

Hillel, D. 1991. *Out of the Earth: Civilization and the Life of the Soil.* Berkeley: University of California Press.

Hillman, G., R. Hedges, A. Moore, S. Colledge, and P. Pettit. 2001. New evidence of Lateglacial cereal cultivation at Abu Hureyra on the Euphrates. *Holocene* 11:383–93.

Köhler-Rollefson, I., and G. O. Rollefson. 1990. The impact of Neolithic subsistence strategies on the environment: The case of ʿAin Ghazal, Jordan. In *Man's Role in the Shaping of the Eastern Mediterranean Landscape,* ed. S. Bottema, G. Entjes-Nieborg, and W. Van Zeist, 3–14. Rotterdam: Balkema.

Lowdermilk, W. C. 1926. Forest destruction and slope denudation in the province of Shansi. *China Journal of Science & Arts* 4:127–35.

Mallory, W. H. 1926. *China: Land of Famine.* Special Publication 6. New York: American Geographical Society.

Mellars, P. 2004. Neanderthals and the modern human colonization of Europe. *Nature* 432:461–65.

Milliman, J. D., Q. Yun-Shan, R. Mei-E, and Y. Saito. 1987. Man's influence on the erosion and transport of sediment by Asian rivers: The Yellow River (Huanghe) example. *Journal of Geology* 95:751–62.

Moore, A. M. T., and G. C. Hillman. 1992. The Pleistocene to Holocene transition and human economy in Southwest Asia: The impact of the Younger Dryas. *American Antiquity* 57:482–94.

Ponting, C. 1993. *A Green History of the World: The Environment and the Collapse of Great Civilizations.* New York: Penguin Books.

Pringle, H. 1998. Neolithic agriculture: The slow birth of agriculture. *Science* 282:1446.

Roberts, N. 1991. Late Quaternary geomorphological change and the origins of agriculture in south central Turkey. *Geoarchaeology* 6:1–26.

Said, R. 1993. *The River Nile: Geology, Hydrology and Utilization.* Oxford: Pergamon Press.

Sarnthein, M. 1978. Sand deserts during glacial maximum and climatic optimum. *Nature* 272:43–46.

Stanley, D. J., and A. G. Warne. 1993. Sea level and initiation of Predynastic culture in the Nile delta. *Nature* 363:435–38.

Wallace, M. 1883. *Egypt and the Egyptian Question.* London: Macmillan.

Westing, A. H. 1981. A note on how many humans that have ever lived. *BioScience* 31:523–24.

Wright, H. E., Jr. 1961. Late Pleistocene climate of Europe: A review. *Geological Society of America Bulletin* 72:933–84.

———. 1976. The environmental setting for plant domestication in the Near East. *Science* 194:385–89.

Zeder, M. A., and B. Hesse. 2000. The initial domestication of goats (*Capra hircus*) in the Zagros Mountains 10,000 years ago. *Science* 287:2254–57.

4. GRAVEYARD OF EMPIRES

Abrams, E. M., and D. J. Rue. 1988. The causes and consequences of deforestation among the prehistoric Maya. *Human Ecology* 16:377–95.

Agriculture in all ages, no.2. 1855. *DeBow's Review* 19:713–17.

Barker, G. 1981. *Landscape and Society: Prehistoric Central Italy.* London: Academic Press.

———. 1985. Agricultural organisation in classical Cyrenaica: the potential of subsistence and survey data. In *Cyrenaica in Antiquity,* ed. G. Barker, J. Lloyd, and J. Reynolds, 121–34. Society for Libyan Studies Occasional Papers 1, BAR International Series 236. Oxford.

Beach, T. 1998. Soil catenas, tropical deforestation, and ancient and contemporary soil erosion in the Petén, Guatemala. *Physical Geography* 19:378–404.

Beach, T., N. Dunning, S. Luzzadder-Beach, D. E. Cook, and J. Lohse. 2006. Impacts of the ancient Maya on soils and soil erosion in the central Maya Lowlands. *Catena* 65:166–78.

Beach, T., N. Dunning, S. Luzzadder-Beach, and V. Scarborough. 2003. Depression soils in the lowland tropics of Northwestern Belize: Anthropogenic and natural origins. In *The Lowland Maya Area: Three Millennia at the Human-Wildland Interface,* ed. A. Gómez-Pompa, M. F. Allen, S. L. Fedick, and J. J. Jiménez-Osornio, 139–74. Binghamton, NY: Food Products Press.

Beach, T., S. Luzzadder-Beach, N. Dunning, J. Hageman, and J. Lohse. 2002. Upland agriculture in the Maya Lowlands: Ancient Maya soil conservation in northwestern Belize. *Geographical Review* 92:372–97.

Betancourt, J., and T. R. Van Devender. 1981. Holocene vegetation in Chaco Canyon. *Science* 214:656–58.

Borowski, O. 1987. *Agriculture in Iron Age Israel.* Winona Lake, IN: Eisenbrauns.

Braund, D. 1985. The social and economic context of the Roman annexation of Cyrenaica. In *Cyrenaica in Antiquity,* 319–25.

Brown, A. G., and K. E. Barber. 1985. Late Holocene Paleoecology and sedimentary history of a small lowland catchment in Central England. *Quaternary Research* 24:87–102.

Cascio, E. L. 1999. The population of Roman Italy in town and country. In *Reconstructing Past Population Trends in Mediterranean Europe (3000 BC–AD 1800),* ed. J. Binfliff and K. Sbonias, 161–71. Oxford: Oxbow Books.

Cook, S. F. 1949. Soil erosion and population in Central Mexico. *Ibero-Americana* 34:1–86.

Cordell, L. 2000. Aftermath of chaos in the Pueblo Southwest. In *Environmental Disaster and the Archaeology of Human Response,* ed. G. Bawden and R. M. Reycraft, 179–93. Maxwell Museum of Anthropology, Anthropological Papers 7. Albuquerque: University of New Mexico.

Dale, T., and V. G. Carter. 1955. *Topsoil and Civilization.* Norman: University of Oklahoma Press.

Deevy, E. S., D. S. Rice, P. M. Rice, H. H. Vaughan, M. Brenner, and M. S. Flannery. 1979. Mayan urbanism: Impact on a tropical karst environment. *Science* 206:298–306.

Dunning, N. P., and T. Beach. 1994. Soil erosion, slope management, and ancient terracing in the Maya Lowlands. *Latin American Antiquity* 5:51–69.

Fuchs, M., A. Lang, and G. A. Wagner. 2004. The history of Holocene soil erosion in the Philous Basin, NE Peloponnese, Greece, based on optical dating. *Holocene* 14:334–45.

Hall, S. A. 1977. Late Quaternary sedimentation and paleoecologic history of Chaco Canyon, New Mexico. *Geological Society of America Bulletin* 88: 1593–1618.

Halstead, P. 1992. Agriculture in the Bronze Age Agean: Towards a model of Palatial economy. In *Agriculture in Ancient Greece,* ed. B. Wells, 105–16. Proceedings of the Seventh International Symposium at the Swedish Institute at Athens, May 16–17, 1990, Svenska Institutet i Athen, Stockholm.

Harris, D. R., and C. Vita-Finzi. 1968. Kokkinopilos—A Greek badland, *The Geographical Journal* 134:537–46.

Heine, K. 2003. Paleopedological evidence of human-induced environmental change in the Puebla-Tlaxcala area (Mexico) during the last 3,500 years. *Revista Mexicana de Ciencias Geológicas* 20:235–44.

Hughes, J. D. 1975. *Ecology in Ancient Civilizations.* Albuquerque: University of New Mexico Press.

Isager, S., and J. E. Skydsgaard. 1992. *Ancient Greek Agriculture: An Introduction.* London: Routledge.

Judson, S. 1963. Erosion and deposition of Italian stream valleys during historic time. *Science* 140:898–99.

————. 1968. Erosion rates near Rome, Italy. *Science* 160:1444–46.

Lespez, L. 2003. Geomorphic responses to long-term landuse changes in Eastern Macedonia (Greece). *Catena* 51:181–208.

Lowdermilk, W. C. 1953. *Conquest of the Land Through 7,000 Years.* U.S. Department of Agriculture, Soil Conservation Service, Agriculture Information Bulletin 99. Washington, DC: GPO.

Marsh, G. P. 1864. *Man and Nature; or, Physical Geography as Modified by Human Action.* New York: Charles Scribner.

McAuliffe, J. R., P. C. Sundt, A. Valiente-Banuet, A. Casas, and J. L. Viveros. 2001. Pre-columbian soil erosion, persistent ecological changes, and collapse of a subsistence agricultural economy in the semi-arid Tehuacán Valley, Mexico's 'Cradle of Maise.' *Journal of Arid Environments* 47:47–75.

McNeill, J. R., and V. Winiwarter. 2004. Breaking the sod: Humankind, history, and soil. *Science* 304:1627–29.

Meijer, F. 1993. Cicero and the costs of the Republican grain laws. In *De Agricultura: In Memoriam Pieter Willem De Neeve (1945–1990),* ed. H. Sancisi-Weerdenburg, R. J. van der Spek, H. C. Teitler, and H. T. Wallinga, 153–63. Dutch Monographs on Ancient History and Archaeology 10. Amsterdam: J. C. Gieben.

Metcalfe, S. E., F. A. Street-Perrott, R. A. Perrott, and D. D. Harkness. 1991. Palaeolimnology of the Upper Lerma Basin, Central Mexico: a record of climatic change and anthropogenic disturbance since 11600 yr BP. *Journal of Paleolimnology* 5:197–218.

O'Hara, S. L., F. A. Street-Perrott, and T. P. Burt. 1993. Accelerated soil erosion around a Mexican highland lake caused by prehispanic agriculture. *Nature* 362:48–51.

Piperno, D. R., M. B. Bush, and P. A. Colinvaux. 1991. Paleoecological perspectives on human adaptation in Central Panama. II The Holocene. *Geoarchaeology* 6:227–50.

Ponting, C. 1993. *A Green History of the World: The Environment and the Collapse of Great Civilizations.* New York: Penguin Books.

Pope, K. O., and T. H. van Andel. 1984. Late Quaternary alluviation and soil formation in the Southern Argolid: its history, causes and archaeological implications. *Journal of Archaeological Science* 11:281–306.

Runnels, C. 2000. Anthropogenic soil erosion in prehistoric Greece: The contribution of regional surveys to the archaeology of environmental disruptions and human response. In *Environmental Disaster and the Archaeology of Human Response*, ed. R. M. Reycraft and G. Bawden, 11–20. Maxwell Museum of Anthropology, Anthropological Papers 7. Albuquerque: University of New Mexico.

Runnels, C. N. 1995. Environmental degradation in Ancient Greece. *Scientific American* 272:96–99.

Sandor, J. A., and N. S. Eash. 1991. Significance of ancient agricultural soils for long-term agronomic studies and sustainable agriculture research. *Agronomy Journal* 83:29–37.

Simkhovitch, V. G. 1916. Rome's fall reconsidered. *Political Science Quarterly* 31:201–43.

Spurr, M. S. 1986. *Arable Cultivation in Roman Italy c.200 B.C.–c.A.D. 100.* Journal of Roman Studies Monographs 3. London: Society for the Promotion of Roman Studies.

Stephens, J. L. 1843. *Incidents of Travel in Yucatán.* Norman: University of Oklahoma Press, 1962.

Thompson, R., G. M. Turner, M. Stiller, and A. Kaufman. 1985. Near East paleomagnetic secular variation recorded in sediments from the Sea of Galillee (Lake Kinneret). *Quaternary Research* 23:175–88.

Turner, B. L., II, P. Klepeis, and L. C. Schneider. 2003. Three millennia in the Southern Yucatán Peninsula: Implications for occupancy, use, and carrying capacity. In *The Lowland Maya Area*, 361–87.

Van Andel, T. H., E. Zangger, and A. Demitrack. 1990. Land use and soil erosion in prehistoric and historical Greece. *Journal of Field Archaeology* 17:379–96.

Vita-Finzi, C. 1969. *The Mediterranean Valleys: Geological Changes in Historical Times.* Cambridge: Cambridge University Press.

White, K. D. 1970. *Roman Farming.* Ithaca: Cornell University Press.

———. 1973. Roman agricultural writers I: Varro and his predecessors. In *Von Den Anfängen Roms bis zum Ausgang Der Republik*, 3:439–97. Aufsteig und Niedergang der Römanischen Welt 1.4. Berlin: Walter de Gruyter.

Williams, M. 2003. *Deforesting the Earth: From Prehistory to Global Crisis.* Chicago: University of Chicago Press.

Zangger, E. 1992. Neolithic to present soil erosion in Greece. In *Past and Present Soil Erosion: Archaeological and Geographical Perspectives*, ed. M. Bell and J. Boardman, 133–47. Oxbow Monograph 22. Oxford: Oxbow Books.

———. 1992. Prehistoric and historic soils in Greece: Assessing the natural resources for agriculture. In *Agriculture in Ancient Greece*, 13–18.

5. LET THEM EAT COLONIES

Bork, H.-R. 1989. Soil erosion during the past millennium in Central Europe and its significance within the geomorphodynamics of the Holocene. In

Landforms and Landform Evolution in West Germany, ed. F. Ahnert, 121–31. *Catena* Suppl. no. 15.

Brown, J. C. 1876. *Reboisement in France: Or, Records of the Replanting of the Alps, the Cevennes, and the Pyrenees with Trees, Herbage, and Brush, with a View to Arresting and Preventing the Destructive Effects of Torrents.* London: Henry S. King.

Clark, G. 1991. Yields per acre in English agriculture, 1250–1860: evidence from labour inputs, *Economic History Review* 44:445–60.

———. 1992. The economics of exhaustion, the Postan Thesis, and the Agricultural Revolution. *Journal of Economic History* 52:61–84.

Cohen, J. E. 1995. *How Many People Can the Earth Support?* New York: W. W. Norton.

De Castro, J. 1952. *The Geography of Hunger.* Boston: Little, Brown.

Dearing, J. A., K. Alström, A. Bergman, J. Regnell, and P. Sandgren. 1990. Recent and long-term records of soil erosion from southern Sweden. In *Soil Erosion on Agricultural Land,* ed. J. Boardman, I. D. L. Foster, and J. A. Dearing, 173–91. New York: John Wiley and Sons.

Dearing, J. A., H. Håkansson, B. Liedberg-Jönsson, A. Persson, S. Skansjö, D. Widholm, and F. El-Daoushy. 1987. Lake sediments used to quantify the erosional response to land use change in southern Sweden. *Oikos* 50:60–78.

Dennell, R. 1978. *Early farming in South Bulgaria from the VI to the III Millennia B.C.* BAR International Series (Supplementary) 45. Oxford.

Edwards, K. J., and K. M. Rowntree. 1980. Radiocarbon and palaeoenvironmental evidence for changing rates of erosion at a Flandrian stage site in Scotland. In *Timescales in Geomorphology,* ed. R. A. Cullingford, D. A. Davidson, and J. Lewin, 207–23. Chichester: John Wiley and Sons.

Evans, R. 1990. Soil erosion: Its impact on the English and Welsh landscape since woodland clearance. In *Soil Erosion on Agricultural Land,* 231–54.

Evelyn, J. 1679. *Terra, a Philosophical Essay of Earth.* London: Printed for John Martyn, Printer to the Royal Society.

Godwin, W. 1793. *An Enquiry concerning Political Justice and Its Influence on General Virtue and Happiness.* Vol. 2. London: Robinson.

Hutton, J. 1795. *Theory of the Earth, with Proofs and Illustrations.* Vol. 2. Edinburgh: William Creech.

Hyams, E. 1952. *Soil and Civilization.* London: Thames and Hudson.

Judson, S. 1968. Erosion of the land, or what's happening to our continents? *American Scientist* 56:356–74.

Kalis, A. J., J. Merkt, and J. Wunderlich. 2003. Environmental changes during the Holocene climatic optimum in central Europe—human impact and natural causes. *Quaternary Science Reviews* 22:33–79.

Lane, C. 1980. The development of pastures and meadows during the sixteenth and seventeenth centuries. *Agricultural Review* 28:18–30.

Lang, A. 2003. Phases of soil erosion-derived colluviation in the loess hills of South Germany. *Catena* 51:209–21.

Lang, A., H.-P. Niller, and M. M. Rind. 2003. Land degradation in Bronze Age Germany: Archaeological, pedological, and chronometrical evidence from a hilltop settlement on the Frauenberg, Niederbayern. *Geoarchaeology* 18:757–78.

Lowdermilk, W. C. 1953. *Conquest of the Land Through 7,000 Years.* U.S. Department of Agriculture, Soil Conservation Service, Agriculture Information Bulletin 99. Washington, DC: GPO.

Lowry, S. T. 2003. The agricultural foundation of the seventeenth-century English oeconomy, *History of Political Economy* 35, Suppl. 1:74–100.

Mäckel, R., R. Schneider, and J. Seidel. 2003. Anthropogenic impact on the landscape of Southern Badenia (Germany) during the Holocene—documented by colluvial and alluvial sediments. *Archaeometry* 45:487–501.

Malthus, T. 1798. *An Essay on the Principle of Population, as It Affects the Future Improvement of Society: with Remarks on the Speculations of Mr. Godwin, M. Condorcet, and Other Writers.* London: J. Johnson.

Markham, G. 1631. *Markhams Farewell to Husbandry; Or, The Enriching of All Sorts of Barren and Sterile Grounds in Our Kingdome, to be as Fruiteful in All Manner of Graine, Pulse, and Grasse, as the Best Grounds Whatsoever.* Printed by Nicholas Okes for John Harison, at the figure of the golden Unicorne in Paternester-row.

Marsh, G. P. 1864. *Man and Nature; or, Physical Geography as Modified by Human Action.* New York: Charles Scribner.

Marx, K. 1867. *Capital: A Critique of Political Economy.* Vol. 1. New York: Vintage Books, 1977.

Melvin, J. 1887. Hutton's views of the vegetable soil or mould, and vegetable and animal life. *Transactions of the Edinburgh Geological Society* 5:468–83.

Morhange, C., F. Blanc, S. Schmitt-Mercury, M. Bourcier, P. Carbonel, C. Oberlin, A. Prone, D. Vivent, and A. Hesnard. 2003. Stratigraphy of late-Holocene deposits of the ancient harbour of Marseilles, southern France. *Holocene* 13:593–604.

Mortimer, J. 1708. *The Whole Art of Husbandry; Or, The Way of Managing and Improving of Land.* London: Printed by F. H. for H. Mortlock at the *Phoenix,* and J. Robinson at the *Golden Lion* in St. Paul's Church-Yard.

Playfair, J. 1802. *Illustrations of the Huttonian Theory of the Earth.* London: Cadell and Davies / Edinburgh: William Creech.

Reclus, E. 1871. *The Earth.* New York: G. P. Putnam and Sons.

Ross, E. B. 1998. *The Malthus Factor: Poverty, Politics and Population in Capitalist Development.* London: Zed Books.

Simkhovitch, V. G. 1913. Hay and history. *Political Science Quarterly* 28:385–403.

Smith, C. D. 1972. Late Neolithic settlement, land-use and Garigue in the Montpellier Region, France. *Man* 7:397–407.

Surell, A. 1870. *A Study of the Torrents in the Department of the Upper Alps.* Trans. A. Gibney. Paris: Dunod.

van de Westeringh, W. 1988. Man-made soils in the Netherlands, especially in sandy areas ("Plaggen soils"). In *Man-Made Soils,* ed. W. Groenman-van Waateringe and M. Robinson, 5–19. Symposia of the Association for Environmental Archaeology 6, BAR International Series 410. Oxford.

Van Hooff, P. P. M., and P. D. Jungerius. 1984. Sediment source and storage in small watersheds of the Keuper marls in Luxembourg, as indicated by soil profile truncation and the deposition of colluvium. *Catena* 11:133–44.

Van Vliet-Lanoë, B., M. Helluin, J. Pellerin, and B. Valadas. 1992. Soil erosion in Western Europe: From the last interglacial to the present. In *Past and Present Soil Erosion: Archaeological and Geographical Perspectives,* ed. M. Bell and J. Boardman, 101–14. Oxbow Monograph 22. Oxford: Oxbow Books.

Whitney, M. 1925. *Soil and Civilization: A Modern Concept of the Soil and the Historical Development of Agriculture.* New York: D. Van Nostrand.

Zangger, E. 1992. Prehistoric and historic soils in Greece: Assessing the natural resources for agriculture. In *Agriculture in Ancient Greece,* ed. B. Wells, 13–19. Proceedings of the Seventh International Symposium at the Swedish Institute at Athens, 16–17 May, 1990. Acta Instituti Atheniensis Regni Sueciae, Series In 4, 42. Stockholm.

Zolitschka, B., K.-E. Behre, and J. Schneider. 2003. Human and climatic impact on the environment as derived from colluvial, fluvial and lacustrine archives—examples from the Bronze Age to the Migration period, Germany. *Quaternary Science Reviews* 22:81–100.

6. WESTWARD HOE

Bagley, W. C., Jr. 1942. *Soil Exhaustion and the Civil War.* Washington, DC: American Council on Public Affairs.

de Beaujour, L. A. F. 1814. *Sketch of the United States of North America.* Trans. W. Walton. London: J. Booth.

Beer, G. L. 1908. *Origins of the British Colonial System, 1578–1660.* New York: Macmillan.

Brissot de Warville, J.-P. 1794. *New Travels in the United States of America, Performed in 1788.* London: J. S. Jordan.

Costa, J. E. 1975. Effects of agriculture on erosion and sedimentation in the Piedmont Province, Maryland. *Geological Society of America Bulletin* 86:1281–86.

Craven, A. O. 1925. *Soil Exhaustion as a Factor in the Agricultural History of Virginia and Maryland, 1606–1860.* University of Illinois Studies in the Social Sciences 13, no. 1. Urbana: University of Illinois.

Craven, J. H. 1833. Letter of John H. Craven. *Farmer's Register* 1:150.

Cronon, W. 1983. *Changes in the Land: Indians, Colonists, and the Ecology of New England.* New York: Hill and Wang.

Eliot, J. 1934. *Essays Upon Field Husbandry in New England and Other Papers, 1748–1762.* Ed. H. J. Carman, R. G. Tugwell, and R. H. True. New York: Columbia University Press.

Glenn, L. C. 1911. *Denudation and Erosion in the Southern Appalachian Region and the Monongahela Basin.* U.S. Geological Survey Professional Paper 72. Washington, DC: GPO.

Gottschalk, L. C. 1945. Effects of soil erosion on navigation in Upper Chesapeake Bay. *Geographical Review* 35:219–38.

Hall, A.R. 1937. *Early Erosion-Control Practices in Virginia.* U.S. Department of Agriculture Miscellaneous Publication 256. Washington, DC: GPO.

Happ, S. C. 1945. Sedimentation in South Carolina Piedmont valleys. *American Journal of Science* 243:113–26.

Hartmann, W. A., and H. H. Wooten. 1935. *Georgia Land Use Problems.* Bulletin 191, Georgia Agricultural Experiment Station.

Hartwell, H., J. Blair, and E. Chilton. 1727. *The Present State of Virginia, and the College.* London: John Wyat.

Hewatt, A. 1779. *An Historical Account of the Rise and Progress of the Colonies of South Carolina and Georgia.* London: A. Donaldson.

Jefferson, T. 1813. Letter to C. W. Peale, April 17, 1813. In *Thomas Jefferson's Garden Book,* annot. E. M. Betts, 509. Philadelphia: American Philosophical Society, 1944.

———. 1894. *The Writings of Thomas Jefferson.* Ed. P. L. Ford. Vol. 3. New York: G. P. Putnam and Sons.

Letter from Alabama. 1833. *Farmer's Register* 1:349.

Lorain, J. 1825. *Nature and Reason Harmonized in the Practice of Husbandry.* Philadelphia: H. C. Carey and L. Lea.

Lyell, C. 1849. *A Second Visit to The United States of North America.* Vol. 2. London: John Murray.

M. N. 1834. On improvement of lands in the central regions of Virginia. *Farmer's Register* 1:585–89.

Mann, C. C. 2002. The real dirt on rainforest fertility. *Science* 297:920–23.

McDonald, A. 1941. *Early American Soil Conservationists.* U.S. Department of Agriculture Miscellaneous Publication 449. Washington, DC: GPO.

Meade, R. H. 1982. Sources, sinks, and storage of river sediment in the Atlantic drainage of the United States. *Journal of Geology* 90:235–52.

Overstreet, W. C., A.M. White, J. W. Whitlow, P. K. Theobald, D. W. Caldwell, and N. P. Cuppels. 1968. *Fluvial monazite deposits in the southeastern United States.* U.S. Geological Survey Professional Paper 568. Washington, DC: GPO.

Pasternack, G. B., G. S. Brush, and W. B. Hilgartner. 2001. Impact of historic land-use change on sediment delivery to a Chesapeake Bay subestuarine delta. *Earth Surface Processes and Landforms* 26:409–27.

Phillips, U. B. 1909. *Plantation and Frontier Documents: 1649–1863*. Vol. 1. Cleveland: Arthur H. Clark.

Ruffin, E. 1832. *An Essay on Calcareous Manures*. Ed. J. C. Sitterson. Cambridge, MA: Harvard University Press, Belknap Press, 1961.

Schoepf, J. D. 1911. *Travels in the Confederation: 1783–1784*. Trans. A. J. Morrison and William J. Campbell. Philadelphia: W. J. Campbell.

Shafer, D. S. 1988. Late Quaternary landscape evolution at Flat Laurel Gap, Blue Ridge Mountains, North Carolina. *Quaternary Research* 30:7–11.

Smith, N. J. H. 1980. Anthrosols and human carrying capacity in Amazonia. *Annals of the Association of American Geographers* 70:553–66.

Stoll, S. 2002. *Larding the Lean Earth: Soil and Society in Nineteenth-Century America*. New York: Hill and Wang.

Taylor, J. 1814. *Arator, Being a Series of Agricultural Essays, Practical and Political*. Columbia: J. M. Carter.

Toulmin, H. 1948. *The Western Country in 1793: Reports on Kentucky and Virginia*. Ed. M. Tinling and G. Davies. San Marino, CA: Henry E. Huntington Library and Art Gallery.

U.S. Congress. Senate. 1850. *Report of the Commissioner of Patents for the Year 1849, part 2, Agriculture*. 31st Congress, 1st sess. Ex. Doc. 15. Washington, DC: GPO.

Washington, G. 1803. *Letters from His Excellency George Washington to Arthur Young, Esq., F.R.S., and Sir John Sinclair, Bart., M.P.: Containing an Account of His Husbandry with His Opinions on Various Questions in Agriculture*. Alexandria, VA: Cottom and Stewart.

————. 1892. *The Writings of George Washington*. Ed. W. C. Ford. Vol. 13. New York: G. P. Putnam and Sons.

White, A. 1910. A briefe relation of the voyage unto Maryland, 1634. In *Narratives of Early Maryland, 1633–1684,* ed. C. C. Hall, 22–45. New York: Charles Scribner.

Wolman, M. G. 1967. A cycle of sedimentation and erosion in urban river channels. *Geografiska Annaler* 49A:385–95.

7. DUST BLOW

Alexander, E. B. 1988. Rates of soil formation: Implications for soil-loss tolerance. *Soil Science* 145:37–45.

Bennett, H. H. 1936. *Soil Conservation and Flood Control*. U.S. Department of Agriculture, Soil Conservation Service, Miscellaneous Publication 11. Washington, DC: GPO.

Bennett, H. H., and W. R. Chapline. 1928. *Soil Erosion, A National Menace.* U.S. Department of Agriculture, Bureau of Chemistry and Soils and Forest Service, Circular 3. Washington, DC: GPO.

Borchert, J. R. 1971. The Dust Bowl in the 1970s. *Annals of the Association of American Geographers* 61:1–22.

Brown, L. R. 1981. World population growth, soil erosion, and food security. *Science* 214:995–1002.

Busacca, A., L. Wagoner, P. Mehringer, and M. Bacon. 1998. Effect of human activity on dustfall: A 1,300-year lake-core record of dust deposition on the Columbia Plateau, Pacific Northwest U.S.A. In *Dust Aerosols, Loess Soils & Global Change,* ed. A. Busacca, 8–11. Publication MISC0190. Pullman: Washington State University.

Catt, J. A. 1988. Loess—its formation, transportation and economic significance. In *Physical and Chemical Weathering in Geochemical Cycles,* ed. A. Lerman, and M. Meybeck, 251:113–42. NATO Advanced Science Institutes Series C: Mathematical and Physical Sciences. Dordrecht: Kluwer Academic.

Clay, J. 2004. *World Agriculture and the Environment.* Washington, DC: Island Press.

Craven, A. O. 1925. *Soil Exhaustion as a Factor in the Agricultural History of Virginia and Maryland, 1606–1860.* University of Illinois Studies in the Social Sciences 13, no. 1. Urbana: University of Illinois.

Davis, R. O. E. 1914. Economic waste from soil erosion. In [1913] *Yearbook of the United States Department of Agriculture,* 207–20. Washington, DC: GPO.

Dazhong, W. 1993. Soil erosion and conservation in China. In *World Soil Erosion and Conservation,* ed. D. Pimentel, 63–85. Cambridge: Cambridge University Press.

Dunne, T., W. E. Dietrich, and M. J. Brunengo. 1978. Recent and past erosion rates in semi-arid Kenya. *Zeitschrift für Geomorphologie, N. F.,* Suppl. 29:130–40.

Hunsberger, B., J. Senior, and S. Carter. 1999. Winds spawn deadly pileups. *Sunday Oregonian,* September 26, A1.

Hurni, H. 1993. Land degradation, famine, and land resource scenarios in Ethiopia. In *World Soil Erosion and Conservation,* 27–61.

Hyams, E. 1952. *Soil and Civilization.* London: Thames and Hudson.

Jacobberger, P. A. 1988. Drought-related changes to geomorphologic processes in central Mali. *Geological Society of America Bulletin* 100:351–61.

Johnson, W. D. 1902. The High Plains and their utilization. In *Twenty-Second Annual Report of the United States Geological Survey,* 637–69. Washington, DC: GPO.

Kaiser, J. 2004. Wounding Earth's fragile skin. *Science* 304:1616–18.

Kaiser, V. G. 1961. Historical land use and erosion in the Palouse—A reappraisal. *Northwest Science* 35:139–53.

Lal, R. 1993. Soil erosion and conservation in West Africa. In *World Soil Erosion and Conservation,* 7–25.

Larson, W. E., F. J. Pierce, and R. H. Dowdy. 1983. The threat of soil erosion to long-term crop production. *Science* 219:458–65.

Le Houérou, H. N. 1996. Climate change, drought and desertification. *Journal of Arid Environments* 34:133–85.

Lowdermilk, W. C. 1935. *Soil Erosion and Its Control in the United States.* U.S. Department of Agriculture, Soil Conservation Service, Miscellaneous Publication 3. Washington, DC: GPO.

———. 1936. *Man-made deserts.* U.S. Department of Agriculture, Soil Conservation Service, Miscellaneous Publication 4.

———. 1941. Conquest of the Land. In *Papers on Soil Conservation, 1936–1941.* U.S. Soil Conservation Service.

Mäckel, R., and D. Walther. 1984, Change of vegetation cover and morphodynamics—a study in applied geomorphology in the semi-arid lands of Northern Kenya, *Zeitschrift für Geomorphologie, N. F.,* Suppl. 51:77–93.

McCool, D. K., J. A. Montgomery, A. J. Busacca, and B. E. Frazier. 1998. Soil degradation by tillage movement. *Advances in GeoEcology* 31:327–32.

Nasrallah, H. A., and R. C. Balling, Jr. 1995. Impact of desertification on temperature trends in the Middle East. *Environmental Monitoring and Assessment* 37:265–71.

National Research Council. Committee on the role of alternative farming methods in modern production agriculture. 1989. *Alternative Agriculture.* Washington, DC: National Academy Press.

Nearing, M. A., F. F. Pruski, and M. R. O'Neal. 2004. Expected climate change impacts on soil erosion rates: A review. *Journal of Soil and Water Conservation* 59:43–50.

Pearce, F. 2001. Desert harvest. *New Scientist* 172:44.

Peng, S., J. Huang, J. E. Sheehy, R. C. Laza, R. M. Visperas, X. Zhong, G. S. Centeno, G. S. Khush, and K. G. Cassman. 2004. Rice yields decline with higher night temperature from global warming. *Proceedings of the National Academy of Sciences of the United States of America* 101:9971–75.

Pimentel, D. 1993. Overview. In *World Soil Erosion and Conservation,* 1–5.

Pimentel, D., J. Allen, A. Beers, L. Guinand, A. Hawkins, R. Linder, P. McLaughlin, B. Meer, D. Musonda, D. Perdue, S. Poisson, R. Salazar, S. Siebert, and K. Stoner. 1993. Soil erosion and agricultural productivity. In *World Soil Erosion and Conservation,* 277–92.

Pimentel, D., C. Harvey, P. Resosudarmo, K. Sinclair, D. Kurz, M. McNair, S. Crist, L. Shpritz, L. Fitton, R. Saffouri, and R. Blair. 1995. Environmental and economic costs of soil erosion and conservation benefits. *Science* 267:1117–23.

Ponting, C. 1993. *A Green History of the World: The Environment and the Collapse of Great Civilizations.* New York: Penguin Books.

Saiko, T. A. 1995. Implications of the disintegration of the former Soviet Union for desertification control. *Environmental Monitoring and Assessment* 37: 289–302.

Sampson, R. N. 1981. *Farmland or Wasteland: A Time to Choose*. Emmaus, PA: Rodale Press.

Schickele, R., J. P. Himmel, and R. M. Hurd. 1935. *Economic Phases of Erosion Control in Southern Iowa and Northern Missouri*. Iowa Agricultural Experiment Station Bulletin 333. Ames: Iowa State College of Agriculture and Mechanic Arts.

Schindler, D. W., and W. F. Donahue. 2006. An impending water crisis in Canada's western prairie provinces. *Proceedings of the National Academy of Sciences* 103:7210–16.

Shaler, N. S. 1891. The origin and nature of soils. In *Papers Accompanying the Annual Report of the Director of the U.S. Geological Survey for the Fiscal Year Ending June 30, 1891*, 211–345. U.S. Geological Survey. Washington, DC: GPO.

———. 1905. *Man and the Earth*. New York: Fox, Duffield.

Swift, J. 1977. Sahelian pastoralists: Underdevelopment, desertification, and famine. *Annual Review of Anthropology* 6:457–78.

Syvitski, J. P. M., C. J. Vörösmarty, A. J. Kettner, and P. Green. 2005. Impact of humans on the flux of terrestrial sediment to the global coastal ocean. *Science* 308:376–80.

Throckmorton, R. I., and L. L. Compton. 1938. Soil erosion by wind. *Report of the Kansas State Board of Agriculture* 56, no. 224-A.

Trimble, S. W., and S. W. Lund. 1982. *Soil Conservation and the Reduction of Erosion and Sedimentation in the Coon Creek Basin, Wisconsin*. U.S. Geological Survey Professional Paper 1234. Washington, DC: GPO.

U.S. Congress. House of Representatives. Great Plains Committee. 1936. *The Future of the Great Plains*, 75th Congress, 1st sess. HD 144. Washington, DC: GPO.

U.S. Department of Agriculture (USDA). 1979. *Erosion in the Palouse: A Summary of the Palouse River Basin Study*. U.S. Department of Agriculture, Soil Conservation Service, Forest Service, and Economics, Statistics, and Cooperative Service.

Wade, N. 1974. Sahelian drought: No victory for Western aid. *Science* 185:234–37.

Wakatsuki, T., and A. Rasyidin. 1992. Rates of weathering and soil formation. *Geoderma* 52:251–63.

Worster, D. 1979. *Dust Bowl: The Southern Plains in the 1930s*. New York: Oxford University Press.

Zonn, I. S. 1995. Desertification in Russia: Problems and solutions (An example in the Republic of Kalmykia-Khalmg Tangch). *Environmental Monitoring and Assessment* 37:347–63.

Appenzeller, T. 2004. The end of cheap oil. *National Geographic* 205 (6): 80–109.

Bennett, H. H. 1947. Soil conservation in the world ahead. *Journal of Soil and Water Conservation* 2:43–50.

Blevins, R. L., R. Lal, J. W. Doran, G. W. Langdale, and W. W. Frye. 1998. Conservation tillage for erosion control and soil quality. In *Advances in Soil and Water Conservation*, ed. F. J. Pierce and W. W. Fry, 51–68. Chelsea, MI: Ann Arbor Press.

Buman, R. A., B. A. Alesii, J. L. Hatfield, and D. L. Karlen. 2004. Profit, yield, and soil quality effects of tillage systems in corn—soybeans. *Journal of Soil and Water Conservation* 59:260–270.

Catt, J. A. 1992. Soil erosion on the Lower Greensand at Woburn Experimental Farm, Bedfordshire—Evidence, history, and causes. In *Past and Present Soil Erosion: Archaeological and Geographical Perspectives,* ed. M. Bell and J. Boardman, 67–76. Oxbow Monograph 22. Oxford: Oxbow Books.

Craswell, E. T. 1993. The management of world soil resources for sustainable agricultural production. In *World Soil Erosion and Conservation,* ed. D. Pimentel, 257–76. Cambridge Studies in Applied Ecology and Resource Management. Cambridge: Cambridge University Press.

Crookes, William. 1900. *The Wheat Problem: Based on Remarks Made in the Presidential Address to the British Association at Bristol in 1898.* New York: G. P. Putnam and Sons.

Drinkwater, L. E., P. Wagoner, and M. Sarrantonio. 1998. Legume-based cropping systems have reduced carbon and nitrogen losses. *Nature* 396: 262–65.

Egan, T. 2004. Big farms reap two harvests with subsidies a bumper crop. *New York Times,* December 26, 2004, 1, 28.

Fan, T., B. A. Stewart, W. A. Payne, W. Yong, J. Luo, and Y. Gao. 2005. Long-term fertilizer and water availability effects on cereal yield and soil chemical properties in Northwest China. *Soil Science Society of America Journal* 69:842–55.

Faulkner, E. H. 1943. *Plowman's Folly.* New York: Grosset and Dunlap.

Hall, A. D. 1917. *The Book of the Rothamsted Experiments.* 2nd ed. Rev. E. J. Russell. New York: E. P. Dutton.

Hilgard, E. W. 1860. *Report on the Geology and Agriculture of the State of Mississippi.* Jackson: E. Barksdale.

Hooke, R. L. 1999. Spatial distribution of human geomorphic activity in the United States: Comparison with rivers. *Earth Surface Processes and Landforms* 24:687–92.

Howard, A. 1940. *An Agricultural Testament.* London: Oxford University Press.

Jackson, W. 2002. Farming in nature's image: Natural systems agriculture. In *The Fatal Harvest Reader: The Tragedy of Industrial Agriculture,* ed. A. Kimbrell, 65–75. Washington, DC: Island Press.

———. 2002. Natural systems agriculture: a truly radical alternative. *Agriculture, Ecosystems and Environment* 88:111–17.

Jenny, H. 1961. "E. W. Hilgard and the Birth of Modern Soil Science." *Agrochimica,* ser. 3 (Pisa).

Johnston, A. E., and G. E. G. Mattingly. 1976. Experiments on the continuous growth of arable crops at Rothamsted and Woburn Experimental Stations: Effects of treatments on crop yields and soil analyses and recent modifications in purpose and design. *Annals of Agronomy* 27:927–56.

Johnson, C. B., and W. C. Moldenhauer. 1979. Effect of chisel versus moldboard plowing on soil erosion by water. *Soil Science Society of America Journal* 43:177–79.

Judson, S. 1968. Erosion of the land, or what's happening to our continents? *American Scientist* 56:356–74.

Lal, R. 2004. Soil carbon sequestration impacts on global climate change and food security. *Science* 304:1623–27.

Lal, R., M. Griffin, J. Apt, L. Lave, and M. G. Morgan. 2004. Managing soil carbon. *Science* 304:39.

Liebig, J. 1843. *Chemistry in Its Application to Agriculture and Physiology.* Ed. from the manuscript of the author by L. Playfair. Philadelphia: James M. Campbell / New York: Saxton and Miles.

Lockeretz, W., G. Shearer, R. Klepper, and S. Sweeney. 1978. Field crop production on organic farms in the Midwest. *Journal of Soil and Water Conservation* 33:130–34.

Mäder, P., A. Fließbach, D. Dubois, L. Gunst, P. Fried, and U. Niggli. 2002. Soil fertility and biodiversity in organic farming. *Science* 296:1694–97.

Mallory, W. H. 1926. *China: Land of Famine.* Special Publication 6. New York: American Geographical Society.

Matson, P. A., W. J. Parton, A. G. Power, and M. J. Swift. 1997. Agricultural intensification and ecosystem properties. *Science* 277:504–9.

McNeill, J. R., and V. Winiwarter. 2004. Breaking the sod: Humankind, history, and soil. *Science* 304:1627–29.

Morgan, R. P. C. 1985. Soil degradation and erosion as a result of agricultural practice. In *Geomorphology and Soils,* ed. K. S. Richards, R. R. Arnett, and S. Ellis, 379–95. London: George Allen and Unwin.

Mosier, A. R., K. Syers, and J. R. Freney. 2004. *Agriculture and the Nitrogen Cycle.* Washington, DC: Island Press.

Musgrave, G. W. 1954. Estimating land erosion-sheet erosion. *Association internationale d'Hydrologie scientifique, Assemblée générale de Rome,* 1: 207–15.

Pimentel, D., P. Hepperly, J. Hanson, D. Douds, and R. Seidel. 2005. Environmental, energetic, and economic comparisons of organic and conventional farming systems. *BioScience* 55:573–82.

Reganold, J. 1989. Farming's organic future. *New Scientist* 122:49–52.

Reganold, J. P., L. F. Elliott, and Y. L. Unger. 1987. Long-term effects of organic and conventional farming on soil erosion. *Nature* 330:370–72.

Reganold, J. P., J. D. Glover, P. K. Andrews, and H. R. Hinman. 2001. Sustainability of three apple production systems. *Nature* 410:926–30.

Reganold, J. P., A. S. Palmer, J. C. Lockhart, and A. N. Macgregor. 1993. Soil quality and financial performance of biodynamic and conventional farms in New Zealand. *Science* 260:344–49.

Rosset, P., J. Collins, and F. M. Lappe. 2000. Lessons from the Green Revolution. *Tikkun Magazine* 15 (2): 52–56.

Ruffin, E. 1832. *An Essay on Calcareous Manures.* Ed. J. C. Sitterson. Cambridge, MA: Harvard University Press, Belknap Press, 1961.

Smil, V. 2001. *Enriching the Earth: Fritz Haber, Carl Bosch, and the Transformation of World Food Production.* Cambridge, MA: MIT Press.

Stuiver, M. 1978. Atmospheric carbon dioxide and carbon reservoir changes: Reduction in terrestrial carbon reservoirs since 1850 has resulted in atmospheric carbon dioxide increases. *Science* 199:253–58.

Tanner, C. B., and R. W. Simonson. 1993. Franklin Hiram King—pioneer scientist. *Soil Science Society of America Journal* 57:286–92.

Taylor, R. H. 1930. Commercial fertilizers in South Carolina. *South Atlantic Quarterly* 29:179–89.

Tiessen, H., E. Cuevas, and P. Chacon. 1994. The role of soil organic matter in sustaining soil fertility. *Nature* 371:783–85.

Truman, C. C., D. W. Reeves, J. N. Shaw, A. C. Motta, C. H. Burmester, R. L. Raper, and E. B. Schwab. 2003. Tillage impacts on soil property, runoff, and soil loss variations from a Rhodic Paleudult under simulated rainfall. *Journal of Soil and Water Conservation* 58:258–67.

Ursic, S. J., and F. E. Dendy. 1965. Sediment yields from small watersheds under various land uses and forest covers. *Proceedings of the Federal Inter-Agency Sedimentation Conference, 1963,* 47–52. U.S. Department of Agriculture, Miscellaneous Publication 970. Washington, DC: GPO.

U.S. Department of Agriculture (USDA). 1901. *Exhaustion and Abandonment of Soils: Testimony of Milton Whitney, Chief of Division of Soils, Before The Industrial Commission.* U.S. Department of Agriculture, Report 70. Washington, DC: GPO.

Van Hise, C. R. 1916. *The Conservation of Natural Resources in the United States.* New York: Macmillan.

Whitney, M. 1909. *Soils of the United States.* U.S. Department of Agriculture, Bureau of Soils Bulletin 55. Washington, DC: GPO.

————. 1925. *Soil and Civilization: A Modern Concept of the Soil and the Historical Development of Agriculture.* New York: D. Van Nostrand.

Wilson, D. 2001. *Fateful Harvest: The True Story of a Small Town, a Global Industry, and a Toxic Secret.* New York: HarperCollins.

Wines, R. A. 1985. *Fertilizer in America: From Waste Recycling to Resource Exploitation.* Philadelphia: Temple University Press.

Yoder, D.C., T. L. Cope, J. B. Wills, and H. P. Denton. 2005. No-till transplanting of vegetable and tobacco to reduce erosion and nutrient surface runoff. *Journal of Soil and Water Conservation* 60:68–72.

9. ISLANDS IN TIME

Arnalds, A. 1998. Strategies for soil conservation in Iceland. *Advances in Geo-Ecology* 31:919–25.

Arnalds, O. 2000. The Icelandic 'Rofabard' soil erosion features. *Earth Surface Processes and Landforms* 25:17–28.

Buckland, P., and A. Dugmore. 1991. "If this is a refugium, why are my feet so bloody cold?" The origins of the Icelandic biota in the light of recent research. In *Environmental Change in Iceland: Past and Present,* ed. J. K. Maizels, and C. Caseldine, 107–25. Dordrecht: Kluwer Academic.

Dugmore, A., and P. Buckland. 1991. Tephrochronology and late Holocene soil erosion in South Iceland. In *Environmental Change in Iceland,* 147–59.

Gerrard, A. J. 1985. Soil erosion and landscape stability in southern Iceland: a tephrochronological approach. In *Geomorphology and Soils,* ed. K. S. Richards, R. R. Arnett, and S. Ellis, 78–95. London: George Allen and Unwin.

Gerrard, J. 1991. An assessment of some of the factors involved in recent landscape change in Iceland. In *Environmental Change in Iceland,* 237–53.

Gísladóttir, G. 2001. Ecological disturbance and soil erosion on grazing land in Southwest Iceland. In *Land Degradation,* ed. A. J. Conacher, 109–26. Dordrecht: Kluwer Academic.

Hunt, T. L., and C. P. Lipo. 2006. Late colonization of Easter Island. *Science* 311:1603–6.

Kirch, P. V. 1996. Late Holocene human-induced modifications to a central Polynesian island ecosystem. *Proceedings of the National Academy of Sciences of the United States of America* 93:5296–5300.

————. 1997. Microcosmic histories: Island perspectives on "global" change. *American Anthropologist* 99 (1): 30–42.

Luke, H. 1952. A visit to Easter Island. *Geographical Magazine* 25:298–306.

Mann, D., J. Chase, J. Edwards, W. Beck, R. Reanier, and M. Mass. 2003. Prehistoric destruction of the primeval soils and vegetation of Rapa Nui (Isla de Pascua, Easter Island). In *Easter Island: Scientific Exploration into the World's*

Environmental Problems in Microcosm, ed. J. Loret and J. T. Tancredi, 133–53. Dordrecht: Kluwer Academic / New York: Plenum.

Mieth, A., and H.-R. Bork. 2005. History, origin and extent of soil erosion on Easter Island (Rapa Nui). *Catena* 63:244–60.

Ólafsdóttir, R., and H. J. Guðmundsson. 2002. Holocene land degradation and climatic change in northeastern Iceland. *Holocene* 12:159–67.

Ponting, C. 1993. *A Green History of the World: The Environment and the Collapse of Great Civilizations.* New York: Penguin Books.

Sveinbjarnardóttir, G. 1991. A study of farm abandonment in two regions of Iceland. In *Environmental Change in Iceland,* 161–77.

Williams, J. 1837. *A Narrative of Missionary Enterprises in the South Sea Islands.* London: J. Snow.

Williams, M. 2003. *Deforesting the Earth: From Prehistory to Global Crisis.* Chicago: University of Chicago Press.

IO. LIFE SPAN OF CIVILIZATIONS

Berry, W. 2002. The whole horse. In *The Fatal Harvest Reader: The Tragedy of Industrial Agriculture,* ed. A. Kimbrell, 39–48. Washington, DC: Island Press.

Cassman, K. G. 1999. Ecological intensification of cereal production systems: Yield potential, soil quality, and precision agriculture. *Proceedings of the National Academy of Sciences of the United States of America* 96:5952–59.

Cassman, K. G., S. K. De Datta, D. C. Olk, J. Alcantara, M. Samson, J. Descalsota, and M. Dizon. 1995. Yield decline and the nitrogen economy of long-term experiments on continuous, irrigated rice systems in the tropics. In *Soil Management: Experimental Basis for Sustainability and Environmental Quality,* ed. R. Lal and B. A. Stewart, 181–222. Boca Raton: Lewis Publishers.

Ehrlich, P. R., A. H. Ehrlich, and G. C. Daily. 1993. Food security, population and environment. *Population and Development Review* 19:1–32.

Engels, F. 1844. The myth of overpopulation. In *Marx and Engels on Malthus,* ed. R. L. Meek, trans. D. L. Meek and R. L. Meek, 57–63. London: Lawrence and Wishart, 1953.

Huston, M. 1993. Biological diversity, soils, and economics. *Science* 262:1676–80.

Kaiser, J. 2004. Wounding Earth's fragile skin. *Science* 304:1616–18.

Larson, W. E., F. J Pierce, and R. H. Dowdy. 1983. The threat of soil erosion to long-term crop production. *Science* 219:458–65.

Pimentel, D., J. Allen, A. Beers, L. Guinand, R. Linder, P. McLaughlin, B. Meer, D. Musonda, D. Perdue, S. Poisson, S. Siebert, K. Stoner, R. Salazar, and A. Hawkins. 1987. World agriculture and soil erosion. *BioScience* 37:277–83.

Pimentel, D., C. Harvey, P. Resosudarmo, K. Sinclair, D. Kurz, M. McNair, S. Crist, L. Shpritz, L. Fitton, R. Saffouri, and R. Blair. 1995. Environmental and economic costs of soil erosion and conservation benefits. *Science* 267: 1117–23.

Saunders, I., and A. Young. 1983. Rates of surface processes on slopes, slope retreat and denudation. *Earth Surface Processes and Landforms* 8:473–501.

Smith, A. 1776. *Inquiry into the Nature and Causes of the Wealth of Nations.* London: W. Strahan and T. Cadell.

Tilman, D. 1999. Global environmental impacts of agricultural expansion: The need for sustainable and efficient practices. *Proceedings of the National Academy of Sciences of the United States of America* 96:5995–6000.

Tilman, D., J. Fargione, B. Wolff, C. D'Antonio, A. Dobson, R. Howarth, D. Schindler, W. H. Schlesinger, D. Simberloff, and D. Swackhamer. 2001. Forecasting agriculturally driven global environmental change. *Science* 292: 281–284.

United Nations Development Programme. 1996. *Urban Agriculture: Food, Jobs and Sustainable Cities.* New York.

Vitousek, P. M., H. A. Mooney, J. Lubchenco, and J. M. Melillo. 1997. Human domination of Earth's ecosystems. *Science* 277:494–99.

Wilkinson, B. H. 2005. Humans as geologic agents: A deep-time perspective. *Geology* 33:161–64.

INDEX

Page numbers in italics refer to illustrations.

271

Columella, Lucius Junius Moderatus, 60, 61, 66–67
Commoner, Barry, 207
common fields, 89–91, 99; enclosure and, 90, 92, 94, 96, 98–99
composting: French gardening and, 243; large-scale, 202–3; terra preta soils and, 142–44
Condorcet, marquis de, 106
conservation tillage, 211–13
contour plowing, 125–26, 134, 160, 173, 175
conventional agricultural methods. See agrochemistry; industrial agriculture; plowing
"convertible husbandry," 94
Cook, Captain James, 221, 222
Cook, Sherburne, 77–78
Coon Creek, Wisconsin, 173–74
cotton cultivation, 125, 128
cover cropping, 127, 175, 213
Craven, Avery, 118
Craven, John, 122–23
Cromwell, Oliver, 108
Crookes, Sir William, 195
cropping systems: British soil improvement and, 95; development of crop production and, 35; erosion rate and, 23; profitability of organic vs. conventional systems and, 201. See also cover cropping; crop rotation; monoculture; polyculture
crop rotation: abandonment of, with fertilizer use, 197; in American agriculture, 124, 173; in European agriculture, 91, 94, 99; as innovation, 180; nitrogen levels and, 193; under Roman Empire, 60–61, 73; soil fertility and, 175, 200–201; two-field systems and, 55. See also legume cultivation; polyculture
crop yields. See productivity of soil
Cuba, 227, 230–32
Cultivator, 128
cultural evolution hypothesis, 30
cultura promiscua, 58
cuneiform tablets, 38
cycle of life, 14–16, 202
cytosine, 15

dam building: ammonia manufacture and, 196, 197; Nile River and, 42–43
Darwin, Charles: earthworms and, 9–13, 16
da Vinci, Leonardo, 88
Davy, Humphrey, 183
De agri cultura (Cato), 59
de Beaujour, Félix, 126
de Castro, Josué, 110
Deere, John, 146, 150
de Fontanière, Jonsse, 103
deforestation: in African Sahel, 165–68; Amazon and, 115–17, 244; American Southwest and, 79–80; in Central America, 112–13; Easter Island and, 218, 220, 221–22; erosion in China and, 45; erosion in Europe and, 85, 86–87, 91; in Ethiopia, 169; French Alps and, 101–2; in Iceland, 224–25; in Italy, 57; in Middle East, 73; Yucatan Peninsula, 74–75, 76–77
de Luc, Jean André, 104–5
Denevan, William, 143
de Orellana, Francisco, 143
De re rustica (Varro), 60, 66
desertification: in Africa, 167–70; as global concern, 170; in Soviet Russia, 164, 165
developed countries, agricultural productivity in, 4, 197–98
developing countries: agitation for land reform in, 110–12; labor-intensive farming and, 242, 244, 245, 246; urban farming and, 243–44
DeYoung, Dennis, 215
Diderot, Denis: Encyclopédie, 101
Diocletian, 88
disk harrow, 204–5
disk plows, 151
Djang, Y. S., 181–82
DNA, 15
dogs, 35
Douglas, Stephen, 135
drainage. See water control
Dred Scott decision, 135
drought: in China, 44; Ethiopia and, 169; global warming and, 171; Great Plains and, 147, 148–49, 151–52, 157; in sub-

farm subsidy programs: conventional vs. organic farming and, 207, 209–10; criticsims of, 210; developing countries and, 244, 245; origins of, 157–58; for unsustainable practices, 245

Faulkner, Edward, 202, 203–5

fertilization. *See* agrochemistry; chemical fertilizers; guano; human waste, as fertilizer; manure; nitrogen; organic methods; phosphorus; soil improvement

Fillmore, Millard, 187

Fitzherbert, John: *Book of Surveying,* 92

floodplains: Chinese agriculture and, 43–44; early agriculture and, 36, 41–42; soil fertility and, 84. *See also* silt accumulation

fodder crops, 94

Ford, Henry, 150

fossil fuels, reliance on, 199–200, 237. *See also* chemical fertilizers

fossil soils, 15

France: agricultural practices in, 87, 100, 103–4, 243; soil loss in the Alps in, 101–2, 103

Franklin, Benjamin, 121–22

Frauenberg (German Neolithic site), 86

French gardening, 243

Garden of Eden, 27

genetically engineered crops, 4, 205, 237, 240

geology: Hutton and, 104–6; soil formation and, 18, 19, 192. *See also* rock weathering; soil ecosystems; topography

Georgia, erosion in, *131,* 131–32, 133

Georgia Courier, 128

Germany: colonialism and, 100; early agriculture in, 85–86; nitrate synthesis and, 195–96

Gila National Forest, 80

Gilbert, Joseph Henry, 184

glacial era, 27–28, 145

Glenn, Leonidas Chalmers, 141

global food supply: agrochemistry and, 195; distribution of, 109, 200; grain reserves and, 198, 240; high-yield grains and, 197–98; postwar productivity increases and, 197–98, 239; promotion of small farms and, 246; reliance on North American grain and, 170

global geography, sustainable intensive farming regions in, 23

global warming, 171, 212–13

goats, 35. *See also* overgrazing

Godwin, William: *Political Justice,* 107

government policies: agroecology and, 242, 244, 245–46; commercial fertilizer industry and, 187–88, 195–96; conservation planning and, 148, 152–53, 155, 172, 173, 176; Cuban agriculture and, 231; European famines and, 109–10; French forests and, 103–4; Haitian agriculture and, 229, 230; homesteading and, 147, 152; loss of biodiversity and, 244; mechanization and, 168; politics of information and, 191–92; U.S. tobacco farming and, 119–20. *See also* farm subsidy programs

grain cultivation: climate and, 171; Great Plains and, 148; in Neolithic Europe, 85; in Roman Empire, 60; soil nitrogen and, 185. *See also* wheat cultivation

grain reserves, 198, 240

granite, 18

Grassland Reserve Program, 163

Great Britain: colonial expansion and, 100, 123; common fields in, 89–91, 92, 94, 96, 98–99; depopulation in, 91–92; early agriculture in, 87–88; land improvement in, 92, 94–99; soil formation in, 9–11, 13; Woburn Experimental Farm in, 158–59

Great Plains: concern about erosion of, 148–50; cultivation of, 147–48; drought and, 147, 148–49, 151–52, 157; droughts and, 147, 148, 157; erosion assessments in, 151; loess soil of, 145, 147, 148; prairie ecosystem of, 146, 147, 148

Greek agriculture, ancient, 50–55

nitrogen fertilization *(continued)*
    sion application and, 242; productiv-
    ity and, 184, 191, 193, 196–97, 239
Nobel Peace Prize, 169, 197
North African agriculture, 64–65, 69–73
North Carolina, erosion in, *139,* 140–41
no-till methods: adoption of, 211, 213, 241,
    242; advantages of, 211–13; erosion
    and, 24, 212; organic farming move-
    ment and, 205–7, 211–13
nutrients in soil, 18, 119, 189; availability of,
    191–92, 205; dependence of life on,
    14–15; discovery of, 183–84

oasis hypothesis, 30
Ocean Island, 187, 188
Oddson, Gisli, 225
O horizon, 21
Oklahoma Territory, 146
organic intensive farming: long-run advan-
    tages of, 205–10; origins of, 202–5
organic matter: erosion rate and, 20–21, 23;
    floodplains and, 41; healthy soil and,
    104, 201, 204–5; methods for reten-
    tion of, 202–5, 242; O horizon and,
    21; terra preta soil and, 142–44
organic methods: adoption of, 241, 242;
    Cuba and, 231–32; soil building and,
    208–9
outsourcing of food production, 110, 125;
    industrialization in Europe and,
    110
overgrazing: African Sahel and, 167;
    ancient agriculture and, 36, 55, 65,
    70, 72, 73; cattle in Amazon and, 117;
    Chinese soil loss and, 46; Iceland
    and, 225, 226

Pacific Islands Company, 187
Palissy, Bernard, 93
Panama, 78–79
Parthenon, 53
Pausanias, 62
Peale, C. W., 125
Pennsylvania agriculture, 127, 129, 194. *See
    also* Rodale Institute, Kutztown,
    Pennsylvania
Persian Gulf, silt buildup in, 39

Pertinax, 64
Peru: Colca Valley agriculture in, 80–81;
    island guano deposits and, 185–87
pest control, natural methods of, 207, 242
pesticides: costs of, and profitability, 199;
    genetically engineered crops and, 205;
    soil-dwelling organisms and, 20
El Petén, Guatemala, 76–77
Philippines, 165
Phoenician civilization, 71–72
phosphate mining: South American guano
    islands and, 187–88; in U.S., 193–94
phosphorus, 16, 18, 193; fertilization with,
    and productivity, 239; precision appli-
    cation of, 242
Piedmont region, in southeastern United
    States, *138*
Pierce, Franklin, 187
Pimentel, David, 174–75
plants: affinities with soil types and, 96;
    cycle of life and, 14–16; domestication
    of, 30–31, 32–33, 34; elements needed
    for growth, 16; nitrogen fixing and,
    18; symbiotic soil biota and, 16–17.
    See also vegetation
Plato, 51, 58
Playfair, John, 105–6
plowing: ancient agriculture and, *37, 41;*
    ancient Greece and, 51–52, 53, 54–55;
    ancient Rome and, 58, 61; cautions
    against, 156, 162; conservation tillage
    and, 211–13; as conterproductive,
    203–4, 205, 206; effects of, 180; ero-
    sion in China and, 45–46; mechaniza-
    tion and, 146, 150–51, *151;* medieval
    Europe and, 89, *90, 91;* for soil
    improvement, 61, 95, 124, 149, 211–13.
    See also no-till methods
Plowman's Folly (Faulkner), 203
Poike Peninsula, Easter Island, 220–21
politics: conservation programs and,
    173–74; nomadic cultures and, 166.
    See also colonialism; social factors;
    war
polyculture, 206–7
Pontine Marshes, 58
population growth: control of, 106–8, 167,
    223–24; early spread of agriculture

and, 30, 34–35, 36–37, 42, 47; relation
to productivity, 34–35, 84, 85–86, 93,
99, 106–7, 195, 198–99, 200; Sumerian agriculture and, 39–40. *See also*
boom-and-bust cycle
Pory, John, 118
potassium, 19, 121, 316
potato blight in Ireland, 108–9
Potomac River, 140
poverty, war on, 176
Powell, John Wesley, 147
prairie ecosystems, 146, 147, 152, 206
prehistoric man, 28–31. *See also* ancient civilizations; Bronze Age agriculture,
Old World; Neolithic agriculture,
Old World
Present State of Virginia, The, 121
productivity: population growth and,
34–35, 84, 85–86, 93, 99, 106–7, 195,
198–99, 200
productivity of soil: animal labor and, 35–
36; in developing countries, 198–99;
European soil improvement and, 98–
99; global warming and, 171; long-
term potential and, 235–36, 238–40;
organic methods and, 207–9; rise of
agriculture and, 34–36; small vs. large
farms and, 159; technological innovation and, 180, 239; twentieth-century
increases in, 183, 197–98. *See also*
boom-and-bust cycle
profitability: high costs of mechanization
and, 150, 157, 158, 160, 161; of large vs.
small farms, 159; no-till methods and,
212; of organic vs. conventional systems and, 159, 201, 207–10
Pueblo culture, 79–80

rainfall. *See* climate
Raleigh, Sir Walter, 118
Randolph, T. M., 125–26
Rapa Nui. *See* Easter Island
ratio of soil erosion to soil formation. *See*
balance in soil dynamics
reaper, 146
Reclus, Jean-Jacques-Élisée, 101–2
reforestation, 102, 103, 223
Reformation, and land reform, 92–93

Reganold, John, 208
religions, and earth-life relation, 27
*Report on the Geology and Agriculture of the
State of Mississippi* (1860), 188
resource depletion: economic systems and,
234–35, 237, 245; prehistory and, 34;
profitability calculation and, 209–10;
social collapse in South Pacific and,
217–23; USDA reports and, 192. *See
also* erosion; erosion rates; topsoil loss
Ricardo, David, 106
rice cultivation, 180–81
rills, 20
rivers. *See* silt accumulation
Rives, Alfred, 140
rock phosphate, 184, 193
rock weathering: earthworms and, 11–13;
physical processes and, 17; plants and,
13, 15; soil nutrients and, 18, 193, 194.
See also geology
Rodale Institute, Kutztown, Pennsylvania,
201, 208
Rofabards (Icelandic soil escarpments), 226,
227
Roggeveen, Jakob, 217
Rolfe, John, 118
Roman Empire, 55–67; agricultural practices in, 58–62, 93; boom-and-bust
cycles in Europe and, 88–89; collapse
of, 63, 66–67, 88, 89, 189; the Nile
and, 67–68; slave labor and, 60, 61,
63, 234; soil erosion in, 58, 62–63
Roosevelt, Franklin D., 152
Rothamsted farm estate, 183–84, 201,
207–8
Ruffin, Edmund: *Essay on Calcareous
Manures,* 129–31
Russian agriculture, 109, 163–65
Rutherford, Daniel, 183

salinization of land: alkali soils and,
190–91; irrigation and, 39, 40, 43,
190–91
saltpeter (potassium nitrate), 121
sandy soil, 17, *182*
scale of society, and responsiveness, 224,
237–38
Scotland, 87–88

Sea of Galilee (Lake Kinneret), 72
sediment. *See* sediment studies; silt; silt
    accumulation
sediment studies: ancient Greek agriculture
    and, 51, 52–55; ancient Roman agri-
    culture and, 55–56, 62; Chinese agri-
    culture and, 34; colonial agriculture
    and, 137–38; Easter Island ecology
    and, 218; European agriculture and,
    85–88; Middle Eastern agriculture
    and, 31–33; Native American civiliza-
    tions and, 75, 77, 78
Shaler, Nathaniel Southgate, 148–49
Shansi (Shanxi) province, China, 45
sheep: domestication of, 35; erosion in Ice-
    land and, 225, 226; manuring of fields
    and, 95. *See also* overgrazing
Silicon Valley, 171
silt: as agricultural soil, 17, 244; Chinese
    agriculture and, 43–44, 45; floodplain
    deposits in Egypt and, 41–42
silt accumulation: colonial agriculture and,
    133, 139–40; Italian rivers and, 58, 88;
    New World rivers and, 103; Persian
    Gulf and, 39. *See also* floodplains
Simkhovitch, Vladimir, 66, 89–90
single-crop farming. *See* monoculture
skin color, evolution of, 28–29
Skinner, John, 185
slash-and-burn agriculture: Amazonia and,
    143; Easter Island and, 220; Mangaia
    Island and, 221–22; Mayan civiliza-
    tion and, 74–75
slave labor: American South and, 117, 125,
    128, 130, 134–37; Roman Empire and,
    60, 61, 63, 234
sloping land, erosion of, 20, 21, 156,
    161–63, 228; contour plowing and,
    125–26, *157*; in Europe, 85–88, 86–87,
    91, 98, 101–3; in Greece, 50–51, 52, 53;
    vegetation and, 21. *See also* marginal
    land, cultivation of; plowing
small farms: American plantations and,
    117, 121, 124; ancient Rome and,
    57–58, 67; costs of mechanization
    and, 150–51, 157, 158; Cuba and, 231;
    debt and, 158, 159, 160; labor-based
    size limits and, 146; loss of, 159, 200;

no-till methods and, 213; subdivision
    of, in Haiti, 229–30; sustainable agri-
    culture and, 190; world hunger and,
    200. *See also* labor-intensive agricul-
    ture
Smith, Adam: *The Wealth of Nations*,
    108–9, 235
Smith, Captain John, 118
social class distinctions, 38, 42, 94
social factors: adoption of no-till methods
    and, 213; industrial agrochemistry
    and, 150–51, 207; loss of commons
    and, 94, 99; Malthus's vs. Godwin's
    views and, 106–8; obstacles to agricul-
    tural reform and, 134; scale of society
    and, 224, 237–38; soil conservation
    and, 128, 161, 168; soil productivity
    and, 94, 194. *See also* colonialism;
    politics
soil, as dynamic interface between rock
    and life, 12–13, 104, 106. *See also* bal-
    ance in soil dynamics; soil ecosys-
    tems; soil horizons
soil biota: nitrogen fixing and, 18, 185, 193;
    in soil ecology, 16–20, 202, 205
soil conservation: early American agricul-
    ture and, 125–30; economic benefits
    of, 174–75; effectiveness of, 173–74,
    210–13; erosion control in 1930s U.S.
    and, 141–42; future of human soci-
    eties and, 234; government policies
    and, 148, 152–53, 155, 172, 173, 176;
    lifespan of civilizations and, 6; in
    Mayan civilization, 74, 75; mecha-
    nization and, 160; modern revolution
    based on, 241; obstacles to, in ante-
    bellum South, 134–37; proven tech-
    nologies for, 175; social conventions
    and, 168; social factors and, 128, 161,
    168; terra preta soil in the Amazon
    and, 142–44. *See also* fallowing;
    manure; no-till methods; organic
    methods; terracing
soil-dwelling organisms. *See* earthworms;
    soil biota
soil ecosystems: agroecology and, 192–95,
    202; climate and, 19, 167; eco-lawns
    and, 201–2; organic methods and,

technological innovation *(continued)*
  productivity increases and, 180, 239;
  seventeenth-century Europe an, 94;
  sustainability and, 202–5, 237. *See also*
  chemical fertilizers; mechanization
Tehuacán Valley, Mexico, 78
tenant farming, 67, 244; the Church and,
  92–93; commercial fertilization and,
  188; European agriculture and, 92, 94;
  soil conservation and, 127, 134, *142,*
  161
terracing: in ancient cultures, 55, 72, 73, 75,
  80; in China, 182; erosion prevention
  and, 24, 104, 160, 182; mechanization
  and, 173
terra preta soil, 142–44
territorial expansion: American agriculture
  and, 120–21, 122, 126–27; guano islands and, 187; Roman agriculture
  and, 58–59, 64–65; slavery in new
  American territories and, 135–37.
  See also colonialism
Tertullian (Quintus Septimius Florens Tertullianus), 65–66
Texas, dust storms in, *153*
Thailand, agriculture in, 144
Theodoric, 88
Theophrastus, 51
Tibetian agriculture, 179–80
Tikopia Island, 221, 222–24
Timgad (ancient North African city), 69,
  70
tobacco cultivation, 118–21
topography: British land management
  and, 98; changes in, and worms,
  11–12; erosion rate and, 20–21, 141;
  factors in soil thickness and, 11–14;
  French land management and,
  103–4; role of erosion in shaping,
  104–6; ruins of ancient cultures and,
  49; soil formation and, 18–19. *See
  also* erosion; gullies; sloping land,
  erosion of
topsoil loss: in Amazon region, 115–17; in
  Caribbean islands, 229; China and,
  182; everyday evidence of, 3, 4; global
  tons annually, 4; in Iceland, 226–27;
  Mississipppi River basin and, 4; in

U.S., 151, 153, 155, 160; in Virgina,
  122–23. *See also* erosion rates; sediment studies; soil conservation
topsoil thickness, factors in, 11–14, *14,*
  21–22, 189. *See also* erosion; erosion
  rates
Toulmin, Harry, 123
toxic waste, as fertilizer, 214–15
tractors, 146, 150, 151. *See also* mechanization
trade: Cuban agriculture and, 230; phosphates and, 187–88. *See also* cash
  crops
"tragedy of the commons," 94
transportation costs, 243
tropical rainforests, 244. *See also* Amazon;
  deforestation; slash-and-burn agriculture
Tunisia, 69
Turkey, 165
Turner, Ted, 233
turnip cultivation, 94
two-field systems, 55

uncertainty: overblown warnings and,
  175–76; as smokescreen for vested
  interests, 24
United Fruit Company, 111
United Nations: Food and Agricultural
  Organization, 110, 200, 229, 243;
  UNESCO archaeological surveys,
  65–66
United States: advent of commercial fertilizers in, 185–87, 188; Civil War in, 103,
  134–37; colonial agriculture in,
  119–24; nitrate production and,
  196–97; soil types in, 22–23; USDA
  estimates of topsoil loss in, 149–50,
  173, 174. *See also* American South;
  government policies; Great Plains;
  industrial agriculture; *entries for specific states*
U. S. Agency for International Development, 229
U. S. Bureau of Agricultural Engineering,
  153
U. S. Department of Agriculture (USDA):
  Bureau of Soils, 192–95; Division of

Text: 11.25/13.5 Garamond
Display: Perpetua
Compositor: Sheridan Books, Inc.
Indexer: Marcia Carlson
Cartographer/Illustrator: Bill Nelson
Printer and binder: Sheridan Books, Inc.